ALASKA'S SALMON FISHERIES

Volume 10, Number 3, 1983
ALASKA GEOGRAPHIC®

Jim Rearden, Chief Editor for this issue

The Alaska Geographic Society

To teach many more to better know and use our natural resources

Chief Editor: Robert A. Henning
Assistant Chief Editor: Barbara Olds
Editor: Penny Rennick
Editorial Assistant: Kathy Doogan
Designer: Sandra Harner
Maps and Charts preparation: Jon.Hersh

ABOUT THIS ISSUE: Jim Rearden, Outdoors Editor of *ALASKA*® magazine, wrote the text for this issue of *ALASKA GEOGRAPHIC*®. A resident of Alaska for 33 years, Jim first came to the territory in 1947 to work as a summer fishery patrol agent for the U.S. Fish and Wildlife Service at Chignik. In 1950 he organized the department of wildlife management at the University of Alaska, Fairbanks, where he taught wildlife and fisheries management for four years.

From 1959 through 1969 he was employed by the Alaska Department of Fish and Game as a commercial fisheries management biologist in Cook Inlet.

From 1970 through 1982 he was variously a member of the Alaska Board of Fish and Game, and the Alaska Board of Game. In 1975-76 he was a member of the National Advisory Committee on Oceans and Atmosphere, appointed to that post by President Gerald Ford. He has been affiliated with Alaska Northwest Publishing Company since 1968.

ALASKA GEOGRAPHIC®, ISSN 0361-1353, is published quarterly by The Alaska Geographic Society, Anchorage, Alaska 99509-6057. Second-class postage paid in Edmonds, Washington 98020-3588. Printed in U.S.A. Copyright© 1983 by The Alaska Geographic Society. All rights reserved. Registered trademark: Alaska Geographic. ISSN 0361-1353; Key title Alaska Geographic.

THE ALASKA GEOGRAPHIC SOCIETY is a nonprofit organization exploring new frontiers of knowledge across the lands of the polar rim, learning how other men and other countries live in their Norths, putting the geography book back in the classroom, exploring new methods of teaching and learning — sharing in the excitement of discovery in man's wonderful new world north of 51°16'.

MEMBERS OF THE SOCIETY RECEIVE *ALASKA GEOGRAPHIC*®, a quality magazine which devotes each quarterly issue to monographic in-depth coverage of a northern geographic region or resource-oriented subject.

MEMBERSHIP DUES in The Alaska Geographic Society are $30 per year; $34 to non-U.S. addresses. (Eighty percent of each year's dues is for a one-year subscription to *ALASKA GEOGRAPHIC*®.) Order from The Alaska Geographic Society, Box 4-EEE, Anchorage, Alaska 99509-6057; (907) 274-0521.

MATERIAL SOUGHT: The editors of *ALASKA GEOGRAPHIC*® seek a wide variety of informative material on the lands north of 51°16' on geographic subjects — anything to do with resources and their uses (with heavy emphasis on quality color photography) — from Alaska, Northern Canada, Siberia, Japan — all geographic areas that have a relationship to Alaska in a physical or economic sense. We do not want material done in excessive scientific terminology. A query to the editors is suggested. Payments are made for all material upon publication.

CHANGE OF ADDRESS: The post office does not automatically forward *ALASKA GEOGRAPHIC*® when you move. To ensure continuous service, notify us six weeks before moving. Send us your new address and zip code (and moving date), your old address and zip code, and if possible send a mailing label from a copy of *ALASKA GEOGRAPHIC*®. Send this information to *ALASKA GEOGRAPHIC*® Mailing Offices, 130 Second Avenue South, Edmonds, Washington 98020-3588.

MAILING LISTS: We have begun making our members' names and addresses available to carefully screened publications and companies whose products and activities might be of interest to you. If you would prefer not to receive such mailings, please so advise us, and include your mailing label (or your name and address if label is not available).

Cover — *Francis Caldwell brings a nice king salmon aboard the troller* Donna C. (Donna Caldwell)

*Previous page — **Boats waiting to sell salmon to fish buyers at Bethel, on the Kuskokwim River.*** (James Barker)

Contents

Letnikof Cove Cannery on the Chilkat Peninsula, near Haines. (Matt Donohoe)

Too often, people Outside ask us from Alaska, "How was the salmon season this year?"

There certainly is not just one salmon season. There are many . . . in dozens of districts from Southeast Alaska to the Arctic . . . and there are five different salmon which run upriver to spawn in thousands of streams at widely varying times of the year.

Then there is the type of fishing your uninformed questioner is perhaps thinking about . . . gill netting, seining, or trolling . . . again, for what species, where?

This comprehensive overview of Alaska's salmon and the salmon industry may help the uninformed and perhaps even put it all in better perspective for those who labor in this far flung and ancient contest of man with the forces of nature that provides one of our most significant foods.

"How was the salmon run?" Where? When?

But whatever, whether it was early man in his stone age years clubbing a spawning salmon like a charging bear, or a modern salmon seiner with a million dollar boat, the catch was a measure of survival.

Tomorrow? Of one thing we can be reassured. In our time, the continuing runs of salmon, though diminished, do continue, in those thousands of streams, in different sizes and species, every season, a living promise that as there was a yesterday, so shall there be another tomorrow.

Robert A. Henning

President
The Alaska Geographic Society

The troller Apollo *crossing*
Queen Charlotte Sound.
(Francis Caldwell)

Alaska's Salmon

After an interval of years, and after swimming thousands of miles in the vast and restless North Pacific, they return to Alaska and their home streams with sides like burnished silver, and with plump streamlined bodies of rich mild-flavored flesh. They are in their prime, brimming with vitality as they swiftly swim into the clear northern waters. They seem exuberant, as they leap and swim and school, stubbornly and steadily moving toward their life's goal, which is to spawn and soon thereafter to die in some Alaskan river or lake where years earlier each was hatched from a golden red egg buried deep in the gravel.

A profligate Nature seemed to have had man in mind when the Pacific salmon evolved: an inspired and inventive man could hardly have conjured an animal more suited for our use. Major traits include:

• salmon return to the place of their origin when they mature, arriving annually with the regularity of the rising sun.

• they are easily caught with relatively simple nets or other types of gear.

• few foods are as nourishing and tasty as salmon, which is palatable whether fresh, dried, smoked, salted, canned, or frozen.

• given a chance, Pacific salmon can become exceedingly abundant.

• salmon draw from the sea the wealth that becomes their rich white to ruby red flesh (varies with species) — with virtually no detrimental impact on man.

• in exchange for their existence their needs are minimal. They require an appropriate freshwater spawning area, and three of the five species (chinook, coho, sockeye) need suitable clean freshwater rivers or lakes for their tiny young for one to three years: pink and chum salmon descend immediately after hatching to the sea to start their explosive marine growth.

• in temperate waters these marvelous Pacific salmon lend themselves nicely to artificial rearing. In Alaska artificial rearing to an age for release in the sea (ocean ranching) has

Troll-caught silvers and kings enjoy the reputation of being top quality. (Francis Caldwell)

*A seiner at Columbia Bay,
Prince William Sound.*
(Tom Walker)

Counting tower of the Alaska Department of Fish and Game at Ugashik Lake in the Bristol Bay area. Summer employees, wearing polarized glasses to reduce glare from the water, count salmon as they pass below the tower. (Frank Bird)

anthropologists say that the complex cultures — especially in the relatively gentle climate of Southeastern Alaska — were free to develop.

When Americans learned to preserve salmon in tin cans in the late 1800s, Alaska's salmon became available to the world. For about a century — until the 1980s — canning was the main method of preserving Alaska's salmon for shipping to distant markets. Only by the early 1980s had air transportation and freezing methods made any serious inroads into the salmon canning industry of Alaska.

Salmon are caught in several ways in various parts of the state, and numerous salmon fisheries exist in the state, each with its own characteristics, type of gear, and method of fishing that gear. Through 1959, under federal control, the regulations were generally adopted six months or more in advance of the fishing season, and they were virtually inflexible: any change had to be published in the federal register before it became effective. Once a regulation was in place, there was little chance of modifying it, regardless of its effect on fishermen or on the salmon.

Flexible management commenced with state control in 1960. The system used by the old Board of Fish and Game (1960-1974) and more recently, the Board of Fisheries (1975 to date) has been to set opening dates, areas to be open to fishing and other general regulations, and to direct the Department of Fish and Game to make adjustments in fishing times and areas by field announcement.

proven to be an economical and practical means of augmenting natural stocks for pink and chum salmon at least, with some potential for the other species.

Except in the high Arctic, Alaska's aboriginal culture was often largely based on salmon. Salmon amply nourished the Yup'ik Eskimos of the west coast, many of the Aleuts, and they were a basic food for the Indians who developed the totemic culture in Southeastern Alaska. Because salmon was a plentiful easily-obtained food, Alaska's pre-white cultures did not have to spend full time seeking food as have other primitive cultures: as result,

The late Eugene "Robbie" Robinson nailing up a salmon regulatory marker. Such markers establish closed fishing areas near stream mouths, and in other areas where fishing cannot be permitted. (Jim Rearden, staff)

*Right — **Bristol Bay gold. Cans of sockeye salmon in the Peter Pan cannery at Dillingham, Bristol Bay.*** (Jim Rearden, staff)

*Above — **The patching table in a Bristol Bay cannery, where women ensure that cans are properly filled with sockeye salmon.*** (Freda Shen)

*Right — **Boxes of salmon roe, bound for Japan. During the past decade the once-wasted roe of salmon has become as valuable as the salmon it comes from. Eggs are salted and carefully packed under supervision of Japanese experts.*** (Freda Shen)

Bristol Bay sockeye salmon being loaded aboard a transport plane at Dillingham. Since about 1979, as much as half of Bristol Bay's sockeye salmon have been marketed as fresh or frozen. A decade ago 90% or more was canned. (Jim Rearden, staff)

Edmund Lord's fish processing plant
with fresh chum salmon, Nenana.
(Lael Morgan, staff)

A Confusion Of Names

There are six species of Pacific salmon. All are commercially valuable and all are caught by fishermen around the Pacific rim. Five are found in North America: the sixth, the Japanese cherry salmon (*Oncorhynchus masu*) is found only in Asia.

A multitude of common names are used for the five North American species of Pacific salmon. Local names vary from Alaska to California and each has at least two common names interchangeably used by most coastal Alaskans. Those who are not familiar with the five salmon species are often confused by changes in body color and shape as these salmon mature sexually in fresh water. A fresh-from-the-sea fish that is bright, silvery, streamlined, and fat, has scant resemblance to the emaciated, red, purple, yellow, or black, fungus covered, humpbacked (with the pink and sockeye salmon especially), or green-headed (red or sockeye), snaggle-toothed horrors that these fish become when they near death at the end of their freshwater breeding cycle.

All Pacific salmon display strong homing instinct and return to the streams of their birth. Sockeye and king salmon have the strongest homing instinct, while the other species may wander, occasionaly ascending a stream near that of their birth. All deposit their eggs in the gravel of freshwater streams or lakes. The female digs the redd, or nest, mostly with her tail, and when she deposits her eggs, the male swims near and releases the sperm-carrying milt. The fish then cover the eggs with gravel, fanning and digging with their tails. The eggs hatch in two to four months, depending upon water temperature. All Pacific salmon die after spawning.

Young salmon hatch with the egg sac attached, and generally remain in the gravel until the egg sac is absorbed, then work their way to the surface.

Length of freshwater life varies with the species, and it varies within the same species (except for pink and chum salmon, which descend to sea immediately upon leaving the gravel). Life in the ocean may last from a few months to many years.

Salmon-spawning time at Brooks Falls, between Lake Brooks and Naknek Lake in Katmai National Park and Preserve. (Thomas Klein, reprinted from ALASKA® magazine)

The **king salmon** (*Onchorhynchus tschawytscha*) is the largest of the five Pacific species. Also called chinook (not "shinook"), spring, tyee, quinnat, Sacramento, or blackmouth, it is found from Monterey, California, northward to Kotzebue Sound, Alaska. An adult king averages about 40 inches long, and it may weigh 40 pounds. Statewide average weight is about 18 pounds for commercially caught kings (see the table in this section for average weights in various parts of Alaska). A giant of its kind, a 126-pounder, was caught near Petersburg in 1949. Weights of between 80 and 90 pounds are not unusual.

Adult king salmon have black, often X-shaped, spotting on the back, dorsal fin, and both lobes of the tail. They also have a black pigment along the gum line — which earns them the name "blackmouth." A fresh sea-run king is a powerful, deep-bodied, streamlined,

*Above — **A small king or chinook salmon. Note spotting on back and dorsal fin, as well as on both lobes of tail.*** (Washington State Department of Fisheries, reprinted from Fisheries of the North Pacific)

*Left — **A Cook Inlet king salmon that blundered into a set gill net is held up by fisherman Andy Miner.*** (Tom Walker)

*Left — **Red salmon spawn in an unnamed creek in Katmai National Park and Preserve — part of the Naknek spawning cycle in early July, with the last fish appearing in late October.*** (Will Troyer, reprinted from ALASKA® magazine)

*Below — **King salmon fry being reared in salt water at Halibut Cove Lagoon, lower Cook Inlet, being examined by former commissioner of Fish and Game, James W. Brooks.***
(Jim Rearden, staff)

Filleting king salmon at Cordova. Bright red flesh of these kings is typical.
(Tom Walker)

lovely, animal — silvery on the sides, white on the belly, and bright shiny blue-green on the back.

King salmon may sexually mature any year between their second and seventh years, accounting for the different sizes of returning spawners. A mature three-year-old may weigh less than 4 pounds: a mature seven-year-old may far exceed 50 pounds.

The flesh of king salmon is firm, and a rich red — except for the meat of "white kings," a genetically different strain which occurs mostly in Southeastern Alaska. About 70% of Southeastern kings are red-meated, 30% are white-meated: white kings are not known to be north of the Alaska Peninsula, and the insignificant numbers found in salt water around the rim of the Gulf of Alaska are probably from Southeastern Alaska. Buyers pay less for white-meated kings, although there is no nutritional difference between the two.

Canned king salmon is usually labeled "chinook salmon."

Sockeye salmon (*Oncorhynchus nerka*), along with the king salmon, bring premium prices to fishermen. Also called red or blueback salmon ("red" and "sockeye" are the most commonly used in Alaska), their flesh is a striking brilliant red, an attractive color that helps bring top prices for this firm-fleshed salmon.

Found from the Klamath River, California, to Point Hope, Alaska, adult sea-going sockeye may reach 33 inches in length, and weigh up to 15.5 pounds. Average is 24 inches and 6 to 9 pounds. The statewide average (in 1980) was 5.6 pounds.

At sea sockeyes are metallic greenish blue on the back, with fine black specks. Belly is lighter. The tail is often a translucent green. In fresh water male sockeye salmon may become goldfish red with an olive-green head; females are darker.

Sockeye may spend from one to five years in salt water maturing — and may reach sexual maturity in from two to seven years. They may remain in fresh water from one to four years.

Sockeye salmon are usually marketed under the name sockeye salmon.

16

*Left — **The Peter Pan fisheries tender Baranof,** at the dock in Dillingham, **loaded with fresh sockeye salmon.***
(Jim Rearden, staff)

A load of Bristol Bay sockeye caught in a set net by fisherwomen Cristi Palin (left) and Audrey Rearden.
(Jim Rearden, staff)

17

Silver or coho salmon when sea fresh are dime bright — with small black spots scattered over the back and upper lobe of the tail. (Washington State Department of Fisheries, reprinted from Fisheries of the North Pacific)

Coho or silver salmon (*Oncorhynchus kisutch*) are found from Monterey Bay, California, to Point Hope, Alaska. Mature adults weigh from 6 to 12 pounds, and average length is 29 inches. The 1980 statewide average weight was 7.1 pounds. Maximum weight is more than 30 pounds.

The coho is an explosive, powerful swimmer, usually sturdily built. The caudal peduncle of a silver salmon is usually broad, and a silvery patch appears on the tail.

At sea, adults are metallic blue on the back, dime-bright silver on the sides and belly, and on the tail. Irregular black spots are scattered over the back and usually on the upper lobe of the tail.

Cohos spend one to three years at sea before returning to their home stream to spawn. Most are four years old at that time.

Canned coho salmon are commonly labeled "medium red" salmon, which is a good description of their firm flesh.

Left — "Cowboy Larry" cleaning fresh-caught coho salmon aboard his troller. (Matt Donohoe)

Below — Coho salmon's silver sides usually turn to a burnished copper color after a few days in the stream. (Bob Hagel, reprinted from ALASKA® magazine)

Pink or humpback salmon (*Oncorhynchus gorbuscha*) are found from northern California to the Arctic Ocean and east to the Mackenzie River. This is the smallest of the Pacific salmon, at maturity averaging 16 to 22 inches long, and weighing about 4 pounds. Average weight, statewide, in 1980, was 3.4 pounds. Maximum weight is about 14 pounds.

Adults have large black spots on the back, adipose fin, and on both lobes of the tail. Spots on the tail are oval. Sides are silvery, with very fine scales. Belly is white. Back is steely blue to blue-green.

Spawning adult males develop an elongated and hooked snout, enlarged teeth, and a pronounced hump on the back. The back and sides become dark, with green-brown blotches. Spawning

*Left — **Unloading pink salmon from a seiner, Prince William Sound.** (Jim Rearden, staff)*

*Above — **A sea-fresh pink salmon. After being in fresh water, pinks develop a pronounced hump on the back — giving it the name humpback salmon.** (Washington State Department of Fisheries, reprinted from* Fisheries of the North Pacific*)*

*Right — **This male pink salmon is from the Norton Sound area near the village of Moses Point, from a river near there. The fish was caught during July in spawning stages.*** (Dennis Gretsch)

*Below — **Spawning female pink salmon. Spawning male pinks develop a great hump and a hooked jaw.*** (Tom Walker)

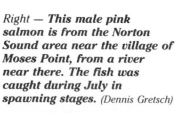

*Right — **Unloading a scow load of pink salmon in Prince William Sound.*** (Jim Rearden, staff)

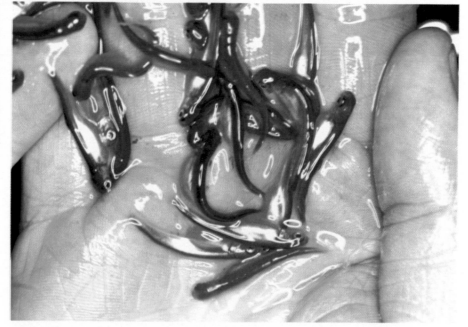

Pink salmon fry, Sheldon Jackson hatchery. (Ernest Manewal)

females do not develop these characteristics as distinctly.

Pink, or humpy salmon, have the simplest and shortest life history of any Pacific salmon. The cycle is two years: spawning adults spawn just two years from the time their parents spawned. Time spent at sea is 14 to 16 months.

Rather soft-fleshed, pink salmon, which do have a pinkish colored meat, must be handled carefully and quickly when caught. Canned humpback salmon are generally labeled "pink salmon."

Chum or dog salmon (*Oncorhynchus keta*) have the largest natural distribution of any of the North American Pacific salmon, ranging from California north to Bering Strait and east to the Mackenzie River.

"Dog salmon," they're often called: there are two guesses as to the derivation of the name. Some believe it resulted because this is the species most often used to feed sled dogs during the years when dogs and sleds provided most of the winter transportation in Interior and Northern Alaska. Others claim the name comes from the hooked snout and vicious-looking doglike protruding teeth of spawning adults.

Chum salmon have weighed 33 pounds, and measured 40 inches long. Average is about 8 pounds and 30 inches in length. In 1980 statewide average weight was 7.4 pounds.

At sea a mature chum is a swift powerful swimmer: chums commonly dive deeply to get beneath fishermen's

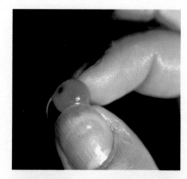

Left — **Chum salmon egg beginning to eye.** *(Matt Donohoe)*

Below — **A spawned-out dog salmon nearing the end of its life cycle.** *(Gerry Atwell, USF&WS, reprinted from* ALASKA GEOGRAPHIC®*)*

Right — **Fishermen aboard the Little Raven, a purse seiner out of Kodiak, haul a catch of silver (coho) salmon aboard.** *(Bernie Kikta, reprinted from* ALASKA® *magazine)*

A sea-fresh chum is as bright and silvery as the other four species of Pacific salmon in Alaska. Note the narrow caudal peduncle (narrowing in front of tail) which is typical.
(Washington State Department of Fisheries, reprinted from Fisheries of the North Pacific*)*

nets. Color of the back is a metallic blue with infrequent dark specking. Pectoral and anal fins, and tail have dark tips. The caudal peduncle is narrow and long, with a strong V between the two lobes of the tail.

Chums spawn after spending two to four years at sea. Including freshwater growth, which is brief, chums are between three to five years old when they return to spawn.

The flesh of chum salmon is light red-dish. It is nutritiously equivalent to the other species of salmon, but the pale coloration is less attractive than the flesh of its relatives, hence it brings less money.

Chum salmon, in the can, is usually labeled, "chum salmon."

Throughout Alaska each fishing district has salmon that are unique. The fine fat king salmon that swim 2,300 miles up the Yukon River are con-

Alaska's Salmon Fisheries
Average Weight in Pounds
For the 1980 commercial salmon season. Based on data from the Alaska Department of Fish and Game.

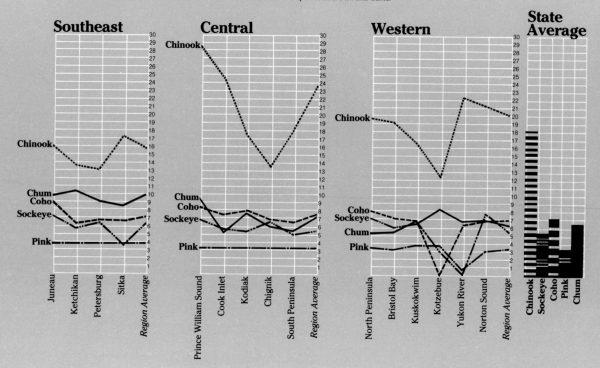

Summary of Data

Area	Chinook	Sockeye	Coho	Pink	Chum
Southeast Alaska					
Juneau	16.0	7.0	8.9	3.8	9.9
Ketchikan	13.6	5.9	6.2	3.8	10.1
Petersburg	13.0	6.3	6.6	3.8	9.2
Sitka	17.1	3.4	6.6	3.8	8.6
Region average	15.9	6.2	7.0	3.8	9.9
Central					
Prince William Sound	28.8	6.8	9.8	3.3	8.2
Cook Inlet	24.4	5.9	5.8	3.3	7.3
Kodiak	17.7	5.4	7.6	3.2	7.6
Chignik	13.7	6.7	6.4	3.3	6.9
South Peninsula	18.2	5.0	5.6	3.1	6.3
Region average	23.8	5.5	7.2	3.2	7.1
Western					
North Peninsula	19.8	5.3	8.0	3.5	7.1
Bristol Bay	19.0	5.5	7.1	3.3	6.0
Kuskokwim	16.6	6.8	6.9	3.7	6.5
Kotzebue	12.9	3.0	.0	3.8	8.6
Yukon River	22.3	.0	6.4	.0	6.8
Norton Sound	21.4	7.8	6.7	3.1	7.1
Region average	20.3	5.5	7.1	3.4	6.7
State average	18.3	5.6	7.1	3.4	7.4

sidered the richest and finest in the state — for the stored oil these fish use as energy to reach spawning grounds. The chum salmon of upper Portland Canal in Southeastern Alaska are probably the largest in the state, at an average of around 20 pounds. The sometimes-huge races of sockeye salmon that ascend the Kenai River are quickly recognized by fishermen.

Because of the varying racial traits of salmon in different fisheries, weights differ considerably from one part of the state to the other. The above table shows the varying size of salmon of the same species in different parts of Alaska.

For comparative purposes, and to look at the current health of the 13 salmon fisheries throughout Alaska, we have concentrated on the 10-year period, 1973-82. A catch table for each area is included at the end of the various sections describing the 13 fisheries. The following table gives the total statewide catch of salmon for the years 1973-1982. Figures are for thousands of fish.

Salmon catch
Statewide

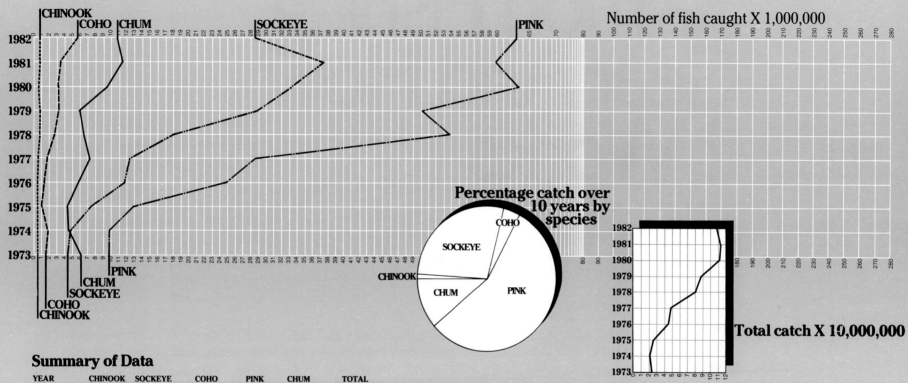

Number of fish caught X 1,000,000

Percentage catch over 10 years by species

Total catch X 10,000,000

Summary of Data

YEAR	CHINOOK	SOCKEYE	COHO	PINK	CHUM	TOTAL
1982	860.2	28,536.8	5,836.8	63,127.8	10,759.9	109,121.4
1981	828.0	37,360.4	3,557.4	59,976.0	11,550.5	113,272.3
1980	675.5	33,308.6	3,135.3	63,282.0	9,611.0	110,012.4
1979	830.4	28,722.7	3,244.5	50,135.6	5,828.8	88,762.0
1978	836.1	18,138.4	2,820.4	53,814.6	6,679.2	82,288.6
1977	621.0	12,460.1	1,815.3	28,586.8	7,328.6	50,811.8
1976	532.7	11,783.3	1,432.3	24,750.8	5,924.6	44,423.7
1975	454.9	7,453.4	1,013.7	12,983.8	4,322.8	26,228.5
1974	557.0	4,878.3	1,858.9	9,858.6	4,728.5	21,881.3
1973	550.6	4,490.2	1,456.7	9,800.5	6,014.9	22,312.9
TOTAL	6,746.4	187,132.1	26,171.3	376,316.5	72,748.8	669,114.9
10-YR. AVERAGE	674.6	18,713.2	2,617.1	37,631.6	7,274.8	66,911.4
% By Species	1.0%	27.9%	3.9%	56.2%	10.9%	

Salmon Fishing Gear And How It Works

Seines catch about half of the salmon in Alaska, statewide, varying with species. Seines take about 85% of the pink salmon, but only about 2% of the king salmon, and about 15% of the sockeye (see graphs). Three types of seines are used — purse seine, hand purse seine, and beach seine. Purse and hand purse seines are fished in a similar manner, from boats: a beach seine is a floating net designed to surround fish which is set from and hauled to the beach.

"A **purse seine** is a floating net designed to surround fish and which can be closed at the bottom by means of a free-running line through one or more rings attached to the lead line." So Alaska defines this great net.

Regulations setting the length and depth of seines vary from district to district throughout Alaska. Minimum lengths vary from 90 to 150 fathoms (One fathom is 6 feet. A long seine is difficult to use easily in a creek where creek robbers may steal salmon from their spawning grounds), and maximum lengths vary from 125 to 250 fathoms, in different areas.

Purse seines can be fished in various ways. Sometimes fishermen see a school of fish and lay their seine out to "wrap up" that school in a quick set. Sometimes a lead (separate net) is used: it is attached to the shore in a place where fish commonly pass. Up to a dozen boats may take turns fishing with the lead: attaching their seine to the

lead for 15 minutes to an hour or more while salmon, which often move with the tide, build up inside the U-shaped nets. When sufficient fish are present the set is completed and the next boat takes its turn at using the lead. At other times a seine without a lead may be held in place in a hook or U-shape, while fish build up in it.

Salmon seine fishermen know salmon are to be found most commonly traveling along beaches, heading for spawning streams. Seine boats may patrol an area, with the skipper inspecting the beaches, watching for the telltale fins ("finners") of a school, or for jumping fish. Some seiners are built with a crow's nest high on the mast equipped with steering and engine con-

A seiner making a set in Salisbury Sound, Southeastern Alaska.
(Matt Donohoe)

*Right — **The seine skiff of the Prince William Sound seiner** Invader, **at the end of a fixed lead, where the skiff man has attached the end of the** Invader's **seine. The** Invader **lays out the seine, and it will "hold a hook" for perhaps 15 minutes, while traveling pink salmon pile up in it. The set will then be completed by the seine skiff holding the end of the net while the** Invader **comes around to pick it up. Another seiner will then hook on to the lead for his turn.***
(Jim Rearden, staff)

*Above — **A hand purse seiner making a set in Seldovia Bay, lower Cook Inlet.***
(Jim Rearden, staff)

trols and the skipper may operate from there: from this height he can look down into the water with polarized glasses to see schools of salmon, often dark against a light bottom, with an occasional silver flash as bright scales reflect light.

As the seiner prowls, searching for fish, the seine lays on the stern, ready to be set or layed out in the water, one end attached to the seine skiff, the other to the seiner. Seine skiffs may be lightweight and outboard powered as they are for hand-pursed seines in Cook Inlet, or, as with most Alaska limit (50-foot) seiners they may be heavy inboard-powered rigs. The seine skiff is towed, snubbed tightly to the stern of the seiner. When fish are near, the skiff man may ride in the skiff, his engine idling, in instant readiness for making a set.

When the boat is in position and the skipper is ready he yells "Let her go!" to the skiff man, who starts his engine if it isn't already running, yanks the lanyard of the quick release mechanism attaching the skiff to the seiner (the "pelican hook"), quickly turns the skiff, and tows the loosely piled seine into the water.

"Let her go!" is a key phrase to salmon seiners: it sets into motion complicated and specialized equipment and skills. It means excitement, hard work — and if fish are caught, money. No

seiner can hear "Let'er go!" without a quickening of pulse.

Some years ago the inebriated crew of an Alaskan salmon seiner found themselves in a courtroom before a magistrate whose name happened to sound like "finner" — another key phrase to a salmon fisherman. One of the happy crew couldn't restrain himself when he heard "finner" as the court was called to order: he pealed out, "A finner? Let 'er go!" All salmon-oriented Alaskans who were present broke up. The magistrate's reaction is unrecorded.

Once a few fathoms of net are in the water the drag of the water plus the forward motion of the roaring seiner

yanks the rest of the seine overboard. The seiner turns a great arc, laying the seine out either around a school of salmon, or forming a U-shape, facing the net into the tide along a beach where salmon are traveling so they will swim into the U.

At the same time the seine skiff continues to tow the end of the net, usually helping to shape the U, depending upon the set. Tangles of web splash into the water and appear to be crisscrossed and snarled — but soon the weight of the sinking lead line pulls the webbing free, and the big plastic floats on the cork line (cork hasn't been used for seine floats for years) keep that edge of the seine on the surface, so that it forms

Above — **After most of the fish are brailed aboard (as seen in the photo at left — this is the same set), the crew rolls the "money bag" aboard, filled with salmon, and dumps them into the hold too. Kachemak Lady is owned and fished by the Frank Wise family of Homer.**
(Jim Rearden, staff)

29

Left — He's a master caulker, one of the last of his kind. Here Jackie Thompson, in his 70s, who has been caulking boats since he was 11, caulks a seam on a salmon seiner at Seaward Shipyard, Ketchikan. (Matt Donohoe)

Right — Reading the scale of a brailer before it is lifted aboard the tender and dumped. (Jim Rearden, staff)

a sort of fence in the water. The rattle of a seine as it leaps from the stern of a speeding seiner is unforgettable.

A seine may be held open for some time, and the skipper usually signals the skiff man his instructions by arm movements: radios aren't often used because of the roar of the seine boat engine and the scream of a towing jitney (another name for a seine skiff) engine.

When it is time to close his set the skipper signals the skiff man and the skiff and seiner head toward each other, closing the circle with the net. The crew grabs lines to the seine, freeing the seine skiff of the net: the skiff man then may run the skiff to the bow, or to the other side of the seiner, to tow the seiner to keep it in deep water, or to keep it from drifting into the seine. With net billowing in the tide, it's easy to get web in the wheel (propeller) or rudder of the seiner, effectively stopping all movement of the seiner until it is cleared.

The end of the seine is hauled through the power block, a huge hydraulically-powered rubber-coated wheel that draws the net aboard, "drying up" the bunt, or the "money bag" —

the heavy-mesh end where the fish are finally held.

During a set, while fish are in the seine and before it is closed, crew members are busy "plunging" — ramming a long-handled tool that resembles a plumber's friend used for toilet stoppages into the water — again and again, forming a curtain of bubbles and creating a disturbance which frightens the fish into remaining within the net until it is pursed.

With a standard purse net, large metal rings are attached to the lead line. A heavy line through these rings is tightened, "pursing" the net — closing

the bottom so that the salmon cannot swim downward and under the net.

With a hand purse seine, the lead line is quickly pulled aboard the seiner, and the fish are trapped between the lead line and cork line. And then the fish are trapped in the end of the seine.

Sometimes the catch is so huge that the weight of the salmon pulls the corks or net floats under, and some of the salmon escape. When the catch is large the seine skiff is brought to the net, the cork line lifted into it, so the skiff acts as a huge float, preventing the cork line from submerging and allowing fish to escape.

The crew of the Colleen *drying up their purse seine in Cholmondeley Sound on the east coast of Prince of Wales Island, Southeastern Alaska. The salmon in their net, if any, are gradually being forced toward the "money bag" or heavy web end of the net.* (Ken Elverum, reprinted from ALASKA® magazine)

If there is a good catch, a large long-handled dip net (brailing net), is plunged into the dark mass of swimming, splashing, and silvery flashing salmon, and when it is filled with fish the gypsy winch is used to lift it aboard. A purse line (usually a light chain) on the brailer is released, and the salmon cascade, still kicking, to thump into the hold.

If the catch is small, the brailer may not be used: hand purse seine fishermen may instead simply pull and roll the net holding the salmon aboard by hand, dumping the splashing and flipping fish into the hold en masse. A section of the net may be tied off with the flopping fish inside it, and that portion of the net may be lifted aboard. Or, sometimes the net is brought through the power block and the fish roll along the net to wind up in the hold.

Once the fish are aboard, the fish-holding end of the net is brought aboard and stacked, the lines made ready again, jellyfish and other accumulated debris swept or washed overboard, the skiff snubbed to the stern again in readiness for another set, and the process starts all over. A good

In October, at the end of the fishing season in Southeastern Alaska, Whitney-Fidalgo's "learner" seiner — a training vessel for young fishermen — makes its last haul and all seine nets are washed down with fresh water before going into winter storage. One of the learners usually ends up standing under the spray of wash water. (Joe Upton, reprinted from ALASKA® magazine)

Sera Baxter (right) and her children Lynx (left) and Brant, picking silver salmon from a set gill net at Kasitsna Bay, lower Cook Inlet.
(Jim Rearden, staff)

crew under ideal conditions may set a 200-fathom seine, brail aboard 2,000 salmon, clean up, and be ready for another set in 45 minutes to an hour.

Gill nets catch about 48% of Alaska's salmon, statewide. This net works on a different principle than a seine. It is designed to entangle fish in the mesh, and consists of a sheet of webbing hung between the cork (floating) line and the lead (heavy, sinking, lead-weighted) line, and it is fished from the surface of the water. Meshes are precisely of a size that allow salmon

to thrust their heads through the mesh and yet small enough to prevent the fish from backing out: normally the gill covers of the fish catch on the individual strings or meshes, preventing the fish from backing up and escaping — hence "gill net."

This is idealistic. All fish are not precisely the same size. Some small fish may be caught in a gill net by being stuck where they are largest in circumference, just in front of the dorsal fin. Other salmon may be so large that their head may not penetrate a mesh,

but their teeth may become caught in the meshes. A gill net is usually constructed or hung so that there is much slack which helps to entangle a salmon: the meshes are roughly diamond shaped, with the long axis vertical.

A strong swift-swimming fish may become gilled, and continue to swim so strongly that the part of the net he is caught in forms a pouch or bag, and then the fish will manage somehow to twist that, and turn and again catch his head and gills in another nearby part of the gill net. To the inexperienced it

Kenai Packers "tin boats" (which are aluminum) tied up at the dock, Cook Inlet.
(Tom Walker)

looks like an impossible task to quickly remove from the net, without dismembering it, a fish tangled in this manner. Much skill is involved in swiftly removing a salmon from a gill net. Special slip-proof nylon gloves are worn, the fish is usually grasped from the top and middle with the right hand, while the left deftly lifts and slides the strong nylon meshes free.

In Alaska there are **drift gill nets** and **set gill nets** used for catching salmon.

Drift gill nets, which catch about 35% of the salmon, statewide, are set from a boat, normally across the path followed by migrating salmon, and allowed to drift freely, usually with a long line from the net attached to the drifting boat. A single "set" or "drift" may last from a few minutes to the better part of a day, depending upon the weather, the number of fish being caught, and the area. Some drift gill-net fishermen leave their main boat and pick fish from the still-fishing drift net with a skiff, thus allowing the net to fish uninterrupted by pulling it. Tiderips are favorite places for salmon to swim in, and are often favorite spots for drift gill-net fishermen. But tiderips also tend to concentrate driftwood and all kinds of flotsam — which can quickly become entangled in a gill net.

Drift gill-net vessels average around 30 feet statewide (Bristol Bay gill-netters may not be more than 32 feet long), but many fishermen successfully fish salmon drift gill nets from open skiffs powered with outboard motors.

Bob Dolan, a Point Baker fisherman, hangs a leadline as he prepares for the fall gill-netting season. Point Baker is a small fishing village near the north tip of Prince of Wales Island, in Southeastern Alaska.
(Joe Upton, reprinted from ALASKA® magazine)

Steve Smith with his drift gill net lying on the clear waters of Prince William Sound. (Will Troyer)

*Left — **Fish coming aboard a Bristol Bay drift gill-netter.*** (Freda Shen)

*Above — **A brailer net full of fresh-caught Bristol Bay red salmon being lifted from an open, outboard-powered skiff from which a drift gill net was fished.*** (Jim Rearden, staff)

Usually made of aluminum or fiberglass, hydraulically powered gill-net reels are used by most drift gill-net boats in most of the state's drift gill-net fisheries. Salmon are picked from the net as it is reeled aboard. In Bristol Bay, where catches of sockeye salmon tend to be very large, and are made quickly, fishermen prefer to use a power roller during the peak of the season: when a net is judged full of fish, the entire gill net is pulled aboard the drift boat, fish and all, and the fish are picked from the net as it lies piled on the deck.

Set gill nets, which catch about 13% of Alaska's salmon, are staked, anchored, or otherwise fixed in place. Set gill nets are normally fished from the shoreline in a great variety of ways. Some set gill nets are staked into place at low tide on a dry beach. They catch salmon when the tide comes in. When the tide drops, the fishermen may walk to the net to pick the fish out, transporting them with anything from a packboard to a wheelbarrow, or with a tractor, or some sort of four-wheel-drive vehicle.

Other set gill nets have one end staked or anchored to shore, and one end may go dry when the tide drops; others may have both ends anchored in deep water where they never go dry.

Most set nets are fished with 18- or 20-foot skiffs powered with outboard motors. The fishermen commonly live in small cabins or tents on the beach while involved in set netting. Often it is a family operation, and becomes a pleasant summer outing and a way of life much enjoyed by those involved.

Above — **Bristol Bay double-ender sailboats during a fishing closure.**

Right — **Double-ender sailboats fished salmon in Bristol Bay until 1952, when, finally, power fishing boats were allowed. The inefficiency of sail was believed to be a conservation measure.**
(Both photos by Charlie Kroll, reprinted from ALASKA® magazine)

A Bristol Bay set net, arranged at low tide, will fish when the tide comes in. Fishermen may wait until tide drops and net goes dry to fish it — or they may pick fish from the net using a skiff.
(Freda Shen)

All **gill nets** are most effective in silty water, or at night, when they are not visible to swimming salmon. The great salmon gill-net fisheries of Alaska — Bristol Bay, Cook Inlet, the Yukon River, Copper River — all have silty water.

The size of mesh, the number of meshes in depth, and the total length for gill nets, is set by regulation, and varies with each fishing district. Total length of drift gill net that a fisherman may fish may be as much as 300 fathoms (1,800 feet) in parts of Southeastern Alaska, to as little as 50 fathoms in the Yukon River.

Set gill net lengths are generally shorter than those for drift gill nets, and legal lengths of set gill nets in various districts vary from 15 to 200 fathoms.

Generally, gill-net mesh (for both set and drift) of 8 inches or more is used for king salmon. Nets with a mesh of 5¼ inches or so (5⅜ inches in Bristol Bay) are used for sockeye, coho, and chum salmon. Nets with a mesh of about 4½ to 4¾ inches is used for pink salmon. Mesh size is measured from knot to knot of one opening of a gill net, with the mesh closed (stretched measure).

Above — **Picking sockeye salmon from a Bristol Bay set net.** *(Freda Shen)*

Right — **Red and Bunny Beeman picking fish from their set gill net at Kalgin Island, Cook Inlet.** *(Tom Walker)*

Left — *A new gill net is, to a fisherman, a thing of beauty.* (Freda Shen)

Below — **Bristol Bay sockeye coming in over the roller in a gill net as skipper Pete Islieb watches.** (Freda Shen)

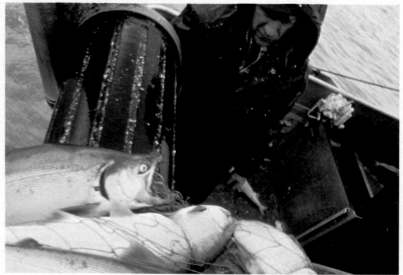

Subsistence fish wheels on the Copper River. *(Tom Walker)*

Fish wheels catch less than 1% of Alaska's salmon. This efficient mechanical contrivance, operated by river current, literally scoops fish from the water. Defined by the Board of Fisheries as ". . . a fixed, rotating device for catching fish which is driven by river current or other means of power," fish wheels are used for commercial salmon fishing only along the upper Yukon River and its drainages. Currently there are about 170 of them in use.

As salmon swim upstream toward their spawning grounds they follow the riverbanks. Fly over the clear-water Kvichak River, for example, during the height of the salmon run there in July and you'll see two black undulating strings of fish — one along each bank of the river. Sometimes the fish swim within arm's length of the bank, or as much as to perhaps 30 feet out, as they steadily move upstream.

The fish wheel is effective because of the salmon's bank-hugging trait. When placed directly over a point where salmon pass, each scoop of the basket has a chance of picking up one or more surprised fish and depositing them in the collecting box of the fish wheel.

Salmon-catching fish wheels came to Alaska from the Sacramento River in California, and the Columbia River in Washington, probably about 1904.

Building a fish wheel, as used on the Yukon River and its tributaries, requires a lot of skill and knowledge. It's mostly ax and chisel work. First a raft is built of logs that are 40 to 50 feet long and from 16 to 20 inches in diameter. An opening roughly 10 feet by 30 feet is left in the raft for the turning fish wheel. The raft is floated into position and tied in place with a several-hundred-foot cable that is attached to a hefty tree upstream. The position of a fish wheel is critical, and some productive fish wheel sites are used year after year. A variance of 6 or 8 feet, sometimes less, can mean the difference between no catch and a large catch. Riverbanks and bottoms constantly change, and new locations are constantly sought and appraised by fish wheel owners.

When the raft is in position a spar or 30-foot pole is attached to the raft and to the beach: it holds the fish wheel in a precise location. Sometimes the spar is used as a bridge for getting from shore to the raft, but most fishermen use a skiff.

Depending upon the site, a lead may be necessary to funnel the fish directly into the path of the turning wheel. Some fish wheel locations are so good a lead isn't used. The lead is simply a heavy wall of poles that is jammed cross-current clear to the bottom next to the bank: when fish encounter the lead they turn and follow it — directly into the arc of the spinning fish wheel.

An axle, made of an 8- or 10-inch pole, or an 8x8 sawed piece about 10 feet long, is prepared with rounded ends that fit into rounded hubs on each side of the opening in the raft. Sometimes these rounded axle ends are greased so they turn freely. Sometimes a short length of pipe is driven onto them. More often soft wire is wrapped around them. The pipe or wire allows easy turning and prevents excessive wear.

The hubs, in which the ends of the axle turn, are adjustable for height. With low water in the river they can be raised, lifting the axle so that the

A water-powered fish wheel on the Yukon River — one of about 170 used by commercial salmon fishermen. (Lael Morgan, staff)

Louis Patterson, of Kenny Lake, with a red salmon he caught on the Copper River. Fish wheels on the Copper are used for subsistence fishing only: commercial use is prohibited. (Tom Walker)

baskets do not hit bottom. When the river level rises, the hubs can be lowered, putting the arc of the baskets close to the bottom where salmon ordinarily swim.

Two baskets are built of green peeled spruce poles that are about three inches in diameter, and 12 to 14 feet long, and are attached to opposite sides of the axle and facing opposite directions. Heavy galvanized chicken fencing wire is used in constructing the baskets.

Two paddles are then built of poles that are somewhat shorter than those of the baskets — perhaps 10 to 12 feet long. They complete the four equadistant segments of the fish wheel — the spokes of the wheel, as it were. Force of the water pushing on the paddles and baskets turns the fish wheel. A chute is built into each basket. As each basket moves out of the water carrying one or more salmon and reaches its high point, the salmon slides

into the chute, which directs the fish into the holding box located at the end of the axle and on the log raft, usually on the river side where the fish can easily be picked up by boat.

Floating driftwood can jam a fish wheel, and it is usually necessary to anchor and tie a string of logs as a shear boom upstream from the fish wheel to divert flotsam away from it.

Fishermen must adjust the fish wheel constantly with changing water levels and current. It can be made to turn faster by increasing the size of the paddles, or slowed by making the paddles smaller. A noisy wheel is said to frighten fish, so most owners try to keep the squeaks down. Constant movement and action of the water can cause the log raft and moving parts of the fish wheel to come apart even though most of them are nailed, bolted, and wired at the various joints. Constant attention is necessary to keep a fish wheel working.

Bill Power's hand-trolling skiff, out of Port Protection, Southeastern Alaska, makes an end-of-the-day drag for silvers in Sumner Strait. (Joe Upton, reprinted from ALASKA® magazine)

Troll gear catches about 2% of Alaska's salmon. A trolling vessel uses a hook and line to catch salmon, mainly chinook and coho, although the other three species occasionally strike the lures or baits used by a troll fisherman.

Trollers are "power trollers" or "hand trollers," depending upon the method used for handling the lines. A power troller has aboard hydraulic or electrically operated gurdies — spools for taking up the line. A hand troller cannot use power for reeling in trolling lines; they are controlled entirely by hand. They may have a hand-powered crank for reeling in the trolling lines. A hand troll vessel may be a sport fishing type of boat, with the fishermen simply using heavy sport tackle with which to troll for salmon.

Salmon trolling vessels are commonly 40 to 50 feet or more long and many are built to weather the open ocean. They carry ice in the holds to allow for a week or more of fishing; in recent years many trollers have installed refrigeration to either chill the catch, or to freeze it.

The **Kathleen Jo,** *a modern troller, owned by Roger Thomas of Port Angeles, Washington, in Alaska's Lituya Bay.* (Francis Caldwell)

Two wooden (most commonly) or aluminum trolling poles, hinged at deck level, rise high above the mast while two more lie back from the bow. As the boat fishes, the poles are dropped outward. From them extend stainless steel lines straight down for 20 or 30 fathoms, held by lead weights of 30 to 65 pounds. Each line may carry multiple leaders with an assortment of brass, chrome, gold, silver, bronze, pearl and variously colored spoons, plugs, hoochies or other artificial bait. Sometimes herring are used.

Trollers fish at 2 or 3 knots (slower and deeper for chinooks, higher and faster for cohos), dragging their lures or baits through the waters where coho or chinook commonly feed. The lines that are trolled are handled by gurdies —power-operated (usually hydraulic) reels or spools that swiftly bring a hooked fish to the boat.

Troll fishermen must have an intimate knowledge of the feeding grounds of salmon. He must know the depths he is fishing, where snags lie that can strip him of his dragging lines, what effect tide changes have in various areas, how currents effect his lures and the fish he seeks.

Commercial salmon trolling is legal only in Southeastern Alaska, and in the southern Gulf of Alaska.

Salmon Harvest by Gear, 1980
Figures supplied by the Alaska Department of Fish & Game

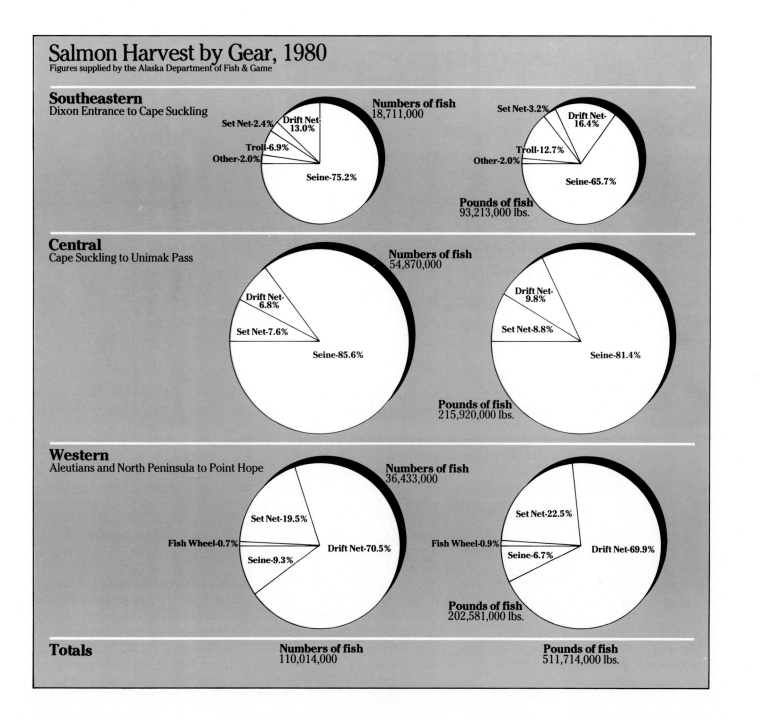

Southeastern
Dixon Entrance to Cape Suckling

Numbers of fish
18,711,000

Set Net-2.4% Drift Net-13.0%
Troll-6.9%
Other-2.0%
Seine-75.2%

Set Net-3.2% Drift Net-16.4%
Troll-12.7%
Other-2.0%
Seine-65.7%

Pounds of fish
93,213,000 lbs.

Central
Cape Suckling to Unimak Pass

Numbers of fish
54,870,000

Drift Net-6.8%
Set Net-7.6%
Seine-85.6%

Drift Net-9.8%
Set Net-8.8%
Seine-81.4%

Pounds of fish
215,920,000 lbs.

Western
Aleutians and North Peninsula to Point Hope

Numbers of fish
36,433,000

Set Net-19.5%
Fish Wheel-0.7%
Drift Net-70.5%
Seine-9.3%

Set Net-22.5%
Fish Wheel-0.9%
Drift Net-69.9%
Seine-6.7%

Pounds of fish
202,581,000 lbs.

Totals

Numbers of fish
110,014,000

Pounds of fish
511,714,000 lbs.

The Fisheries

Coastal Alaska, from Dixon Entrance to Point Hope, includes 13 different salmon management areas, or fisheries. Each has different regulations, and its own state fishery management officials. Within each of these there are multiple salmon fisheries of differing types of gear, species of salmon, and timing of runs. A "salmon fishery" could be considered to be one species of salmon bound for one stream, but we have chosen to consider each of Alaska's 13 major salmon management areas as a "salmon fishery," in a broad brush approach. Each area has its own special geography, people, and "flavor," and catch records are individually kept for each by state fishery managers.

Southeastern Alaska

Southeastern Alaska is fished by purse seines and drift gill nets, as well as hand and power trollers. In addition, there are four floating fish traps at Annette Island, the only remaining salmon fish traps in Alaska.

The Southeast Alaska salmon fleet, like others throughout the state, has become increasingly more efficient and competitive in recent years. High investment in vessels, gear, and in some cases, entry permits, have created an incentive to use each unit of gear to its maximum. Larger high-speed vessels, especially in the drift gill-net fleet, have entered the fishery and

there is a tendency for the fleet to move from district to district to fish on the peaks of certain runs, or to move from districts where extensive closures are in effect.

During the 10 years 1973-82 the Southeast Alaska catch averaged about 15.8 million salmon annually — or about 23.7% of the total statewide catch of salmon (in numbers of fish). Most of these fish were pink salmon (77.5%), for Southeastern Alaska is pink salmon country. There's good reason. This is a watery, island region, flanked by steep down-to-the-water glacier-capped mountains of the mainland.

Southeastern Alaska has 67 islands of more than 2,500 acres, and more than 185 islands that are smaller. It

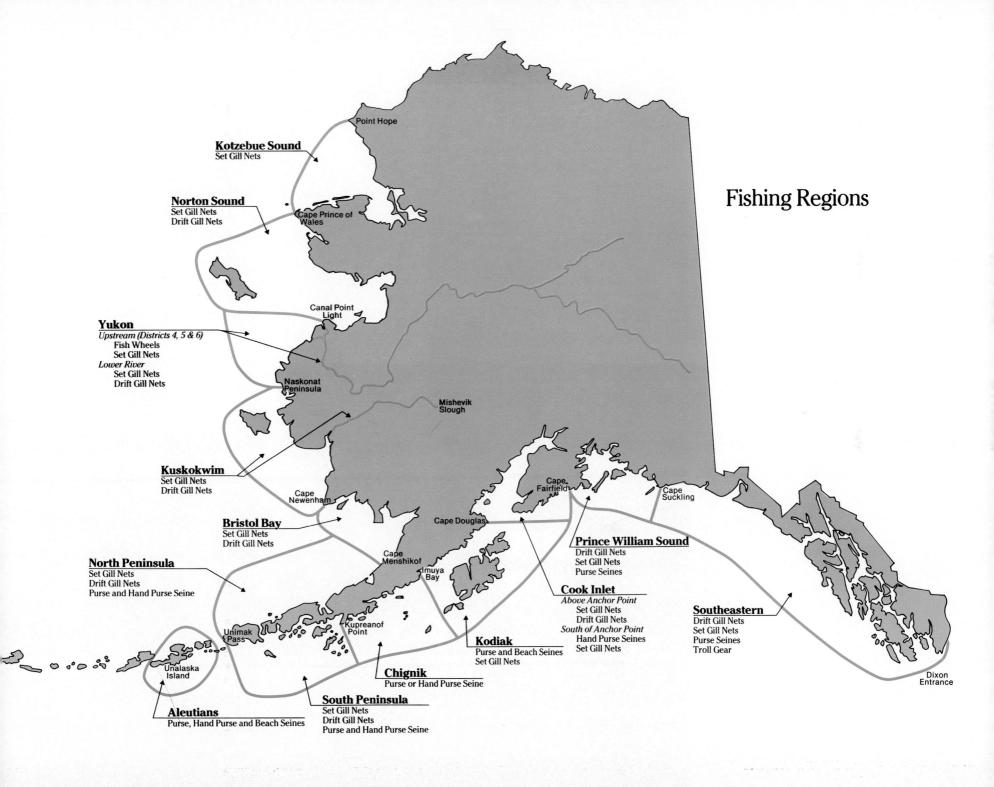

Fishing Regions

Kotzebue Sound
Set Gill Nets

Norton Sound
Set Gill Nets
Drift Gill Nets

Yukon
Upstream (Districts 4, 5 & 6)
Fish Wheels
Set Gill Nets
Lower River
Set Gill Nets
Drift Gill Nets

Kuskokwim
Set Gill Nets
Drift Gill Nets

Bristol Bay
Set Gill Nets
Drift Gill Nets

North Peninsula
Set Gill Nets
Drift Gill Nets
Purse and Hand Purse Seine

Aleutians
Purse, Hand Purse and Beach Seines

South Peninsula
Set Gill Nets
Drift Gill Nets
Purse and Hand Purse Seine

Chignik
Purse or Hand Purse Seine

Kodiak
Purse and Beach Seines
Set Gill Nets

Cook Inlet
Above Anchor Point
Set Gill Nets
Drift Gill Nets
South of Anchor Point
Hand Purse Seines
Set Gill Nets

Prince William Sound
Drift Gill Nets
Set Gill Nets
Purse Seines

Southeastern
Drift Gill Nets
Set Gill Nets
Purse Seines
Troll Gear

Point Hope

Cape Prince of Wales

Canal Point Light

Naskonat Peninsula

Mishevik Slough

Cape Newenham

Cape Menshikof

Imuya Bay

Cape Douglas

Cape Fairfield

Cape Suckling

Unimak Pass

Unalaska Island

Kupreanof Point

Dixon Entrance

Purse seining in Clarence Strait, along the east coast of Prince of Wales Island.
(Dan Kowalski, reprinted from ALASKA® magazine)

stretches nearly 600 miles north and south (Dixon Entrance to Cape Suckling) and the southern half of inland waterways and islands averages 120 miles east to west. There are about 30,000 miles of tidal shoreline — 63% of Alaska's total — between Dixon Entrance and Icy Cape. There are hundreds of sheltered bays, surrounded by conifer-clad, mountainous islands that rise to 3,500 feet. Dozens of deep channels (400 feet or more) wend complex routes among the islands. Most prominent is Chatham Strait, a deep trench 4 to 15 miles wide, and 200 miles long. There are many sheltered bays and channels, but there are also many rough-water runs where winds may whoop along broad and long channels, and where seaworthy boats are lost almost every year.

Southeastern Alaska has a maritime climate of small temperature variations, high humidity, high precipitation, much cloudiness, and little freezing weather. The average temperatures range from the 40s into the 60s (F) in summer, and from the teens to the low 40s in winter. Average annual precipitation is more than 100 inches.

These conditions have produced a lush forest, which is part of the cool northern rain forest that extends along the Pacific coast from northern California to Cook Inlet, Alaska. The Southeast Alaska forests are dominated by western hemlock and Sitka spruce, with smaller numbers of Alaska and red cedar. Major logging shows have denuded many areas in southern Southeastern Alaska, and they are

A troller deckload of coho salmon and a rough sea.
(Matt Donohoe)

A troller running past the can at Frederick Sound and into Petersburg. The Devils Thumb is in the background. (Matt Donohoe)

gradually extending these cuts into northern areas.

Seabirds abound, with many nesting rookeries. Big game includes Sitka black-tailed deer, moose, mountain goat, both black and brown/grizzly bear, wolf, wolverine.

Few roads exist in this region: the Alaska Marine Highway system (ferries) and air travel provide most of the access. Major communities include Juneau, Alaska's capital (pop. 19,483), Ketchikan (7,248), Sitka (7,769), Petersburg (2,800), and Wrangell (2,174) (figures are those within city boundaries; area populations are larger).

Many streams are short and swift as they course through dense coniferous forests. These are most suitable for pink and chum salmon, which require only a suitable place for eggs to be deposited: their young descend to the sea immediately upon hatching. Many coho salmon are also produced in quiet back waters, sloughs, and trickles of Southeastern: 45.2% of the coho salmon caught in Alaska during the 10 years 1973-82 were taken in Southeastern.

Pink salmon: This lovely, wet, rich land has produced one of the world's great pink salmon fisheries, and though all five species of Pacific salmon that are found in North America are here in commercial abundance, it is the overwhelming number of pink salmon that dictates most management policies for the important purse seine fishery, and for which canneries and purse seine fishing boats are largely geared. The gill-net and troll fleets, and some cold

storage plants, are not geared for pink salmon. There are more than 2,000 pink salmon spawning streams or systems in Southeast Alaska. However, the pink salmon runs have the greatest fluctuations in run size of any of the five species. Further, even and odd-year (pinks are strictly two-year fish: returning adults are invariably the progeny of parents which spawned two years previous) pink salmon exhibit different run timing characteristics, and in some cases utilize certain streams to varying degrees.

In decades past Southeastern Alaska historically produced half or more of Alaska's pink salmon: the northern half of Southeastern (roughly north of the latitude of Petersburg) produces about one-third of the total pink catch in the panhandle.

This has changed in recent years. Between 1973 and 1982, Southeast Alaska produced about 32% of the total pink salmon caught in the state (123 million pinks, compared with 376.3 million caught statewide).

Before 1900 the salmon fishery of Southeastern Alaska was limited largely to red salmon, but with increased demand for pinks, the pink fishery developed rapidly, peaking in 1941 with a harvest of 68 million. The annual catch during the developing period averaged more than 30 million.

Then a decline came, and the catch reached a low of 5.5 million in 1960. Recovery under state management (which started in 1960) appeared good at first, then disaster struck with a low of 3.7 million pinks caught in 1975.

Trollers tied up in the Alaska Native Brotherhood Harbor, at Sitka. (Matt Donohoe)

Since then the catch has gradually increased as runs have rebuilt, with an average catch of 12.3 million a year in the 10 years 1973-82.

Extreme cold during winters of the late 1960s and early 1970s most likely caused the low salmon returns of the early 1970s. Plankton, which tiny pink fry depend upon for food in salt water, live and reproduce best in warming waters of spring. If coastal waters aren't warm enough to produce plankton when fry reach the sea (usually about March), mortality can be very high. Plankton abundance is but one of the numerous factors that can affect salmon survival throughout their lives.

Studies indicate there is a general north-south separation of migrating adult salmon: those bound for northern Southeast Alaska streams enter inside waters through Icy Strait, some through Peril Strait, and others through lower Chatham Strait. Those bound for southern streams enter through Sumner Strait and Dixon Entrance. Pink salmon bound for streams on the outside coast generally do not enter inside waters.

Peaks of spawning for early run pinks in both northern and southern Southeast Alaska are from August 10 to 15, the middle run peaks from August

53

15 to September 1, and the late run peaks from September 1 to early October.

Southeastern Alaska's pink salmon have been caught with beach seines, traps, purse seines and gill nets. By the early 1920s purse seines and traps became the primary gear. From 1920 until the early 1950s traps took the bulk of the fish. Trap numbers peaked in 1927 when 575 were used throughout Southeastern Alaska. By 1940 there were only 250 traps in use, which was cut by half by federal fishery managers in 1954.

Traps were abolished by state law in 1959, except for the four at Annette Island owned by Natives with aboriginal rights at Annette Island. Since elimination of most of the traps, seines have regularly caught about 80% of the pink salmon taken in Southeastern Alaska.

Chum salmon are of relatively minor importance in Southeastern Alaska, making up 7.1% of the catch in that part of the state in recent years (1973-82). An average of 1.1 million chums were annually caught in Southeastern during those years, or 15.7% of the statewide total of this species. However, chums are two to three times the size of pink salmon, so numbers of fish cannot directly be compared when value or volume of fish are being considered.

Most chums in Southeastern are normally caught by seine fishermen (around 60%), while drift gill-net fishermen take about 30%.

Chum runs appear in early summer, during summer, and in the fall. Early summer runs appear before, or coincidentally with, very early pink runs. The summer chum run coincides with the main pink run, while the fall chum run is normally quite late — usually after the pink season. One study found that 90% or more of Southeastern Alaska chums are four-year fish, while 10% or less are three-year fish.

Fish Creek, at the head of Portland Canal, is home stream for a unique population of chums that average about 20 pounds at maturity — more than twice the size of most chums found elsewhere in Alaska.

In addition to the chum runs that share most of Southeastern streams with pink salmon, fall run chums enter specific streams after most other Alaskan runs have ended. Starting in early September, these runs extend into early November. Historically these runs have been in great demand both as a commercial fish and as a subsistence species. Rounding out a busy summer with a trip to their island cabins to smoke "fall dogs" has for years been a tradition in Southeastern Alaska.

These late chum runs had been lowered below the levels of commercial harvest in some areas long before statehood, but they have shown signs of rebuilding. Although the harvest is still limited in some areas, the fall fishery is regaining some of its previous significance to the commercial harvest. The Chilkat River, at Haines, on the north end of Lynn Canal, supports a run of unusually large (10 pounds or more) late chums well into November. They spawn in upwellings along the Chilkat River that commonly do not freeze, despite extreme cold.

Sockeye or red salmon are highly important to gill-netters in Southeastern Alaska. Between 1973 and 1982 an annual average of 911,400 sockeye were caught there by fishermen, which is about 4.8% of the total number of sockeye taken statewide. Important sockeye fisheries include those of the Stikine River, Taku Inlet, Port Snettisham, and Chilkat and Chilkoot rivers.

Chinook salmon, highly prized and much sought after throughout Alaska, reach their highest value in Southeastern Alaska, for here more than 90% of them are caught by the elite of Alaska's commercial salmon fishermen — trollers. Southeast Alaska chinooks are virtually all "spring" run: they return to their home streams in April-June, with spawning in mid-summer. Mature fish are caught from April through June as they migrate from outer coastal areas into inside waters.

The instant a freshly caught chinook is lifted aboard a troller it is killed, then it is cleaned, and put on ice — or in a refrigerated hold. Troll fishermen count the worth of their chinook salmon catch in dollars per pound, while gill-netters receive perhaps a third to one fourth the amount paid to trollers for their chinook salmon. Reason? Quality. There is no finer salmon produced in Alaska.

From about 1900, catches of chinook salmon increased in Southeastern

*Opposite — **The salmon troller Uccello, her poles rigged out, fishing off the west coast of Baranof Island, Southeastern Alaska.** (Matt Donohoe)*

A lone troller works the waters off the west coast of Baranof Island. (Matt Donohoe)

Troller George Hicks aboard his vessel, Emma, at Sitka.
(Matt Donohoe)

Alaska until 1919 when about 637,000 of them were caught, mostly by trollers.

The 10-year peak catch was during the years 1931-40, when the average annual catch of chinooks in Southeastern Alaska was about 591,000 fish. This catch has steadily declined and the annual average catch for the years 1973-82 was 316,800, of which more than 90% were caught by trollers. Why the reduced catch? It is probably the result of coastwide overfishing as well as the decline of Columbia River chinook stocks, which is a significant contributor to Southeast Alaska's chinook catch.

Natural chinook salmon runs occur in 33 rivers and streams in Southeastern Alaska. The major producers, with potential runs exceeding 10,000 fish, are the Stikine, Taku, and Alsek rivers, all three of which originate in Canada: these rivers produce about 70% of the total Southeastern Alaska-reared chinook salmon.

Medium producers (500 to 5,000 spawners) include the Keta, Chickamin, Unuk, Harding, Chilkat, Situk, Bradfield, and Wilson-Blossom rivers. These eight streams contribute about 20% of the chinooks reared in Southeastern Alaska. The King Salmon River on the north end of Admiralty Island is the only island stream with significant numbers of spawning chinooks.

A program is currently in progress to rebuild depressed natural Southeast Alaska chinook runs. More restrictive regulations begun in the mid-1970s (most major net fisheries specifically for chinook salmon in Southeastern Alaska have been closed since the mid-1970s) and substantially expanded in 1981 and 1982 (closures of the season, and reducing the chinook catch limit from 320,000 in 1980 to 285,000 in 1981, and 257,000 in 1982) appear to have significantly increased escapements of chinooks to most streams that have been monitored for this species.

The modern salmon troller is capable of fishing for up to two weeks without stopping, and it carries sufficient ice — or has refrigeration — to keep fish until delivery. A few trollers even freeze their fish and deliver when and where they receive the highest price.

Average weight of chinooks taken in the Southeastern Alaska troll fishery has increased in recent years. Legal size limit was 26 inches prior to 1977: changed to 28 inches in 1977.

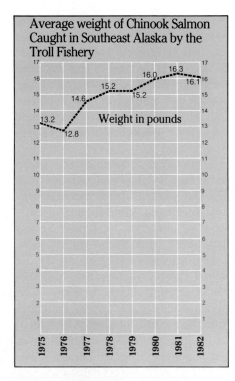

Average weight of Chinook Salmon Caught in Southeast Alaska by the Troll Fishery

Weight in pounds

13.2
12.8
14.6
15.2
15.2
16.0
16.3
16.1

1975 1976 1977 1978 1979 1980 1981 1982

A major problem with chinook salmon is hook mortality — sublegal fish, or "shakers" (that must be "shaken" off the hook: commercially caught kings must be released if they are under 28 inches). An unknown but significant number of chinook salmon die from this cause.

The troll fishery for chinook salmon in Alaska is difficult to manage, for the fish are of mixed stocks: that is, the fish originate from many widely scattered river systems within Alaska and outside of Alaska. The contribution of each of the river systems to the catch is unknown and not regulated.

Coho or silver salmon, husky, explosive in its leaps, spawns in more than 2,000 streams in Southeastern Alaska. Fewer than 100 of these systems have more than 2,000 spawners. The limiting factor is probably the quality of the freshwater rearing habitat rather than availability of spawning gravel — cohos spend one or two years in fresh water after hatching, and before going to sea.

Like most salmon fisheries of Alaska, the Southeastern coho fishery built from about 1900 with an ever-increasing catch until the all-time high was reached in 1951 when 3.3 million were taken. The average annual catch for the 10 years 1973-82 was 1.1 million of which about 62% were caught by trollers, 17% by seiners, and roughly 10% each for set and drift gill-net fishermen. For the past decade and a half the average annual catch has been fairly stable at between 1 and 1.5 million fish.

Southeastern Alaska consistently produces the largest catch of coho salmon of any single fishing area of the state: during the 10 years 1973-82 it produced 45.2% of the cohos caught in Alaska.

Cohos were probably overharvested during the 1920s, 1930s, and 1940s, when peak catches were made. Catches since the late 1950s have been closer to the maximum sustained yield. The 1982 catch of 1.9 million was among the highest in years — a phenomenon noted in almost every fishing district in the state.

Cohos are mostly fish of small streams and tributaries of large mainland rivers, and they live in fresh

A catch of troll-caught king salmon taken by Ray Rodgers with his **Helen A.** *(Matt Donohoe)*

57

Snow on the water? Its sea foam from a storm. A troller works in the distance, off the west coast of Kruzof Island.
(Matt Donohoe)

water for an average of two years after hatching. Spawning and rearing habitat are critical. Clearcut logging to stream banks, dragging trees in streams, road building, pollution from oil, pulp, sewage and other sources, and increasing urbanization all have detrimental impacts on the cohos of Southeastern Alaska.

Some rivers of Southeastern Alaska have runs in July and October — a bimodal pattern. These systems may not have been bimodal runs in early years — intensive midsummer fishing may have created the present situation.

Like king salmon, coho stocks are mixed, with salmon from different streams being taken together. This makes it difficult for fishery managers to make closures of seasons and areas that will protect spawners for all areas. Logging and other economically important land uses that alter habitat of coho-producing streams is difficult to control, and is probably the major threat to Southeast Alaska's coho salmon.

The Yakutat-Yakataga subdistrict of Southeastern Alaska extends from Cape Suckling to Cape Fairweather, where the major salmon fishery consists of set-netters in estuaries or various rivers. All five species of salmon are caught by these fishermen, although sockeye, coho, and pink salmon are the major species landed. Troll fishermen commonly fish in Yakutat Bay and in the offshore waters.

The catch for the Yakutat-Yakataga district in total numbers of fish is about 2% to 3% of the total catch for Southeastern Alaska.

Salmon catch for
Southeastern Alaska

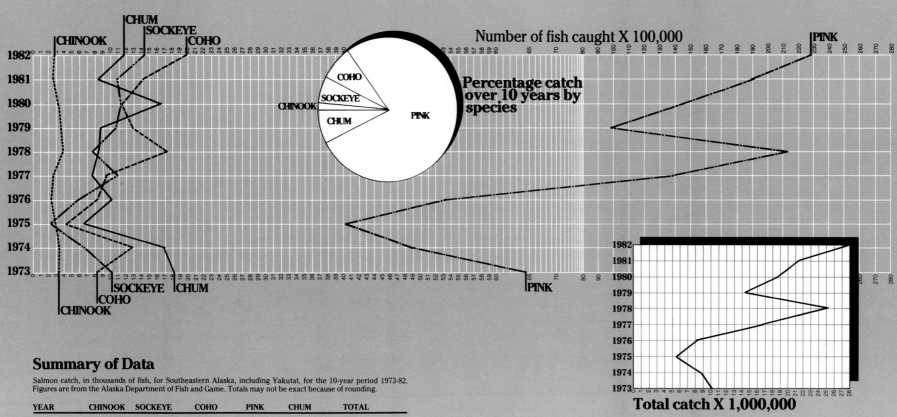

Number of fish caught X 100,000

Percentage catch over 10 years by species

Total catch X 1,000,000

Summary of Data

Salmon catch, in thousands of fish, for Southeastern Alaska, including Yakutat, for the 10-year period 1973-82. Figures are from the Alaska Department of Fish and Game. Totals may not be exact because of rounding.

YEAR	CHINOOK	SOCKEYE	COHO	PINK	CHUM	TOTAL
1982	285.9	1,428.2	1,991.1	22,868.7	1,187.4	27,761.3
1981	271.9	1,079.6	1,407.1	18,967.9	849.5	22,576.1
1980	323.2	1,120.4	1,136.7	14,478.3	1,651.2	18,709.9
1979	367.6	1,073.9	1,284.6	10,977.9	888.3	14,592.3
1978	401.4	788.3	1,714.5	21,243.4	869.0	25,016.6
1977	285.2	1,085.1	944.8	13,843.6	738.7	16,897.4
1976	241.8	595.3	823.7	5,329.6	1,030.9	8,021.2
1975	300.7	245.2	427.4	4,026.5	686.6	5,686.4
1974	346.6	687.4	1,278.2	4,888.8	1,682.6	8,883.5
1973	343.6	1,011.5	836.3	6,455.2	1,832.2	10,478.2
TOTAL	3,168.0	9,114.9	11,844.4	123,079.9	11,416.4	158,622.9
AVERAGE	3,168.0	911.4	1,184.4	12,307.9	1,141.6	15,862.2
% Each Species 10 Yrs.	1.9%	5.7%	7.4%	77.5%	7.1%	
% Of Statewide Catch For Species, 10 Yrs.	46.9%	4.8%	45.2%	32.0%	15.7%	ALL: 23.7%

Limited entry permits held for Southeastern Alaska (September 1982) included: 421 purse seine, 484 drift gill net, 2,150 hand troll, and 484 power troll.

Prince William Sound

Prince William Sound is fished by purse seines, drift gill nets and set gill nets. As in all other areas of the state salmon fishermen of Prince William Sound have become increasingly more efficient in recent years. High investment in vessels, gear, and entry permits have created an incentive to use each boat and each net as effectively as possible. The fishing vessels are faster and can carry more fish and synthetic fiber nets are stronger and far more efficient than those of earlier decades.

During the 10 years 1973-82 Prince William Sound fishermen caught an annual average of 10.5 million salmon of five species, or 15.7% of the state's total salmon catch. Most of these salmon (83.5%) were pink salmon (an annual average of 8.8 million, 1973-82) for, like Southeastern Alaska, Prince William Sound is pink salmon country. In recent years (1973-82) Prince William Sound produced 23.3% of the pink salmon caught in Alaska.

Next numerous are sockeye salmon, with an annual average of 796,800 caught during the years 1973-82, 4.2% of the statewide sockeye catch. Also, during these years, an annual average of 643,190 chum salmon (8.8% of the statewide catch in recent years), 270,360 coho salmon (10.3% of the statewide catch), and 25,140 chinook

A small Prince William Sound seiner at anchor, awaiting a tide and time to go fishing. (Jim Rearden, staff)

Chum and pink salmon in fresh water, as they approach spawning, near Valdez. Dead, spawned-out fish lay on the bottom. The schooled fish have yet to break up into pairs. When paired and actively spawning, salmon scatter out. (Tom Walker)

salmon (3.7% of the statewide catch) were caught in Prince William Sound.

Prince William Sound proper is administratively divided into nine major fishery districts, six for the management of the purse seine fishery for pink and chum salmon, and three smaller districts for the management of sockeye salmon runs which are caught by set gill nets, drift gill nets, and purse seines.

In addition, the Copper River and Bering River districts are included in the Prince William Sound management area; both are restricted to drift gill-net fishing, mainly for sockeye and coho salmon.

Purse seiners, which catch most of the fish in the sound, fish all Prince William Sound districts, except Eshamy, usually beginning in early or mid-July (late July in some years), depending upon the strength of early pink salmon runs, and usually fish into the first or second week of August.

Mountain-surrounded, island-dotted, sheltered Prince William Sound is roughly 60 miles north and south, and 125 miles east and west at the northernmost point in the Gulf of Alaska. The sound is a huge bay containing many coves, bays, harbors, sheltered fjords, narrows and straits, lagoons and islands. Islands range down from 50-mile-long Montague, which lies between the violent Gulf of Alaska and the sound. The Copper River basin, which contributes most of the sockeye and chinook salmon to the Prince

William Sound catch, has long cold winters, and short hot summers. The great valley from which this turbulent giant flows has many clear streams and lakes which are ideal for spawning red and chinook salmon, where fry can grow and feed for the year or three they spend in fresh water, after which they descend the swift stream to the sea.

This is a region of great tectonic activity, with frequent severe earthquakes. One of the world's strongest recorded earthquakes occurred here on March 27, 1964, with shock waves over a 500-mile-wide area. It registered between 8.4 and 8.6 on the Richter scale with the primary epicenter in northern Prince William Sound. A vertical shift ranging from 32 feet of uplift to 6 feet of subsidence occurred in various pink and chum salmon streams of Prince William Sound, with a profound effect on salmon runs to those streams: the pink salmon catch for the sound decreased sharply immediately following the great earthquake.

The most pronounced effect of this uplift and subsidence was in the intertidal area where 35% to 77% of pink salmon spawning had occurred. Stream banks and beds became unstable and surface flows dropped or disappeared in the uplifted intertidal area. Channels were exposed to salt water and increased sedimentation in areas of subsidence. It seems that somehow nature can compensate for even great disasters: decreased numbers of spawning salmon occurred in streams where changes were observed, and increased spawning occurred on unchanged streams.

Between 1967 and 1971 the Alaska Department of Fish and Game attempted to improve Prince William Sound salmon spawning areas damaged by the great earthquake, using state and available federal money. The goal was to stabilize pink and chum salmon streams that had been uplifted or had subsided. Mechanical renovation of such streams included: 1. emplacement of drop structures, 2. channel extension, 3. bank stabilization, 4. channel excavation, and 5. construction of water deflectors on nine high value streams.

The work had to be done between the times when immature salmon left the stream and when spawning adults returned — from about early April until July. The ADF&G vessel *Shad* with a 24' x 60' barge, a tractor, a loader-backhoe, and a crew with hand tools performed the work.

In addition to mechanically improving streams, pink salmon spawners were also fenced into streams. Incubation boxes were used in an attempt to renew lost pink, chum, and sockeye salmon stocks, using eggs from streams with surviving stocks. Adult sockeye spawners were transplanted to Solf Lake from Eshamy Lake.

After several years of field work there was no evidence that the production of pink salmon was increasing, so mechanical renovation was stopped in 1972, and only limited restocking work continued.

Incubation boxes in a few locations produced good survival of pink salmon eggs, while several others failed, due primarily to inadequate water supply.

The natural evolution of streams eased the problem, and the bulldozing and other work speeded the process. Within about a decade of the big earthquake spawning areas had returned to near-normal, and the salmon fishery of Prince William Sound has essentially recovered from what, at the time, appeared to be severe damage.

Volcanism shaped many of the peaks that surround the sound, and scientists keep a wary eye on many of these apparently dormant mountains. Glaciers provide source water for many of the mainland streams, including the great Copper River, one of Alaska's largest.

High peaks of the Chugach Mountains on the east and north, and the great Kenai Mountains on the west, form three walls around Prince William Sound. Moisture-laden air from the Gulf of Alaska dumps vast amounts of rain and snow on the steep and high mainland mountains to create a typical wet and windy maritime climate, with abundant snow and rain, varying between different parts of the sound. During fall and winter winds of the gulf commonly blow from 30-50 MPH, occasionally reaching 100 MPH.

Cordova's average annual temperature is about 38° F, with precipitation of about 92 inches: Valdez, farther from the sea, is cooler, averaging 36° F, with 59 inches of precipitation. For real precipitation, there is Whittier, at the base of the high weather-blocking

*Opposite — **Sheep Bay, Prince William Sound, with an anchored fishing boat.*** *(Jim Rearden, staff)*

*Below — **A box of fresh-caught coho salmon is lifted from a salmon boat.*** *(Matt Donohoe)*

*Below — **Crewmen pile the purse seiner of the** Invader **during a set in Prince William Sound. Net is being pulled aboard by the hydraulically powered block at the top of this photo.*** *(Jim Rearden, staff)*

*Right — **Harry Hamilton, skipper of the seiner** Invader, **at Point Elrington, Prince William Sound, waiting his turn at the lead.*** *(Jim Rearden, staff)*

Chugach Range, with an average of 175 inches annually. These are the three major ports in Prince William Sound.

The pink salmon fishery of the sound has produced more than 1 billion pounds since 1896. Peak year for pinks in Prince William Sound was 1982 when about 20.3 million of the spotted, rich-fleshed swimmers were caught as they returned to their home spawning streams.

An abrupt increase in catch of Prince William Sound pink salmon, from an average of about 3 million, occurred in 1979 when 15.6 million were caught: each year since has seen catches above any previous year's back to 1910. The 1980 catch was 14.2 million, in 1981 it was 20.5 million, and the 1982 catch was 20.3 million. Since 1910, the pink catch of the sound has exceeded 10

million only eight times: four of those have been since 1979.

The abundant short streams and the high rainfall of Prince William Sound closely resemble those of Southeastern Alaska, and, as in Southeastern, pinks bound for widely distributed streams arrive at capes and passes and points in large schools as they reenter the sound. Fishermen meet them, and large numbers of the schooled fish are taken at these well-known concentration points or "cape fisheries."

Sockeye salmon. The average catch of sockeye salmon from Prince William Sound (including the Copper and Bering rivers) averaged 796,800 annually during the 10 years 1973-82. Most of these fish were caught in the Copper River fishery.

The Copper River District opens for the fishing of sockeye salmon in mid-May, the earliest salmon fishery in Prince William Sound, and one of the earliest in the state. The Copper River fishery averages a catch of about 625,000 sockeye salmon annually. The 1982 catch of 1.2 million was the highest recorded in recent years.

The dry catch statistics don't show it, but the Copper River drift gill-net fishery is probably the most dangerous salmon fishery in Alaska. Fishing takes place off the mouth of the great, muddy, rapids-filled Copper River, which flows for 250 miles from the north side of the Wrangell Mountains through the Chugach Range into the Gulf of Alaska east of Prince William Sound.

Upon their return as approximately 7-pound ruby-fleshed, silver-sided, fat, vigorous adults, eager fishermen, with 24- to 30-foot mostly fiberglass bow-pickers (the net is layed and retrieved from the bow), with inboard-outboard power, brave the surging seas just outside the breakers where they lay out their 150 fathoms of drift gill net to catch these prime fish. During the 1950s through the 1970s Copper River flats fishermen used beamy, shallow-drift, rugged, outboard-powered "Cordova skiffs," with a tiny forward cabin and a gill net roller astern. Sudden storms and cranky slow-starting engines that allowed boats to drift into the violent surf both claimed many lives over the years in this rich fishery.

The Bering River District, adja-

A drift gill-netter speeds across Prince William Sound. (Frank Bird)

cent to the Copper River, is a drift fishery mostly for coho and sockeye salmon: fishing starts in mid-June and may continue as late as mid-September. Coho salmon is the major species in the Bering River District, with average catches of around 60,000 annually.

The earliest salmon fishery in the sound proper is that of the Coghill-Unakwik district, which starts in late June and normally ends about mid-July for drift gill nets. Purse seine fishing in these districts coincide with drift gill-net fishing, but lasts beyond the mid-July gill net closing date in order to harvest later runs of pink and chum salmon.

A surprise run of sockeye hit the Coghill district in 1982. The 10-year average catch prior to 1982 was 108,000; but the 1982 catch was 942,000 sockeye, averaging 6.5 pounds. Further, at the Coghill weir 180,000 sockeye were counted into spawning areas: goal for that area was 50,000 to 60,000. For every parent spawner, about 40 fish returned in 1982, an almost unheard-of survival rate.

The Eshamy district, where there is a late red salmon run, usually begins in early July and extends into September. Purse seines fishing in the Southwestern district in July and August catch about 30% of the Eshamy reds before they reach the gill-net fishery.

Chum salmon, like the pink salmon, thrive in the short stream environment of Prince William Sound proper. Over the years chum catches have been less prone to great variations in run size: it is a relatively stable fishery. In the early 1970s the maximum sustained yield (MSY) for the chum fishery for the sound was estimated to be about 605,000 fish annually, based upon the catches of the years 1924-48.

The average catch of chums for the years 1973-82 was 643,190 — reasonably close to the estimated MSY.

Coho salmon are caught in Prince William Sound proper mostly incidentally to pink and chum salmon in the purse seine fishery. The Copper and Bering rivers gill-net fisheries produce far more cohos. The cohos are incidental to the sockeye during the early season, but later the catch is almost all cohos. The estimated maximum sustained yield for the Copper-Bering coho fishery is about 207,000 fish annually.

The total coho catch for Prince William Sound, including that of the Copper-Bering rivers, averaged 270,360 fish annually from 1973-82, with catches that ranged from 76,000 to 615,000 — the latter a record breaker in 1982.

King Salmon of Prince William Sound are mostly caught incidentally to the sockeye in the Copper River fishery. During the 10 years 1973-82 the annual average catch (for all Prince William Sound, including the Copper and Bering rivers) was 25,140 kings. The 1982 catch of kings, following the trend of other species for the sound, was also a record breaker, with 49,200 fish netted. The previous high catch of kings was 32,800 caught in 1976.

Pink salmon piled in the hold of a cannery tender bound for the cannery, Prince William Sound. (Jim Rearden, staff)

Salmon catch for
Prince William Sound

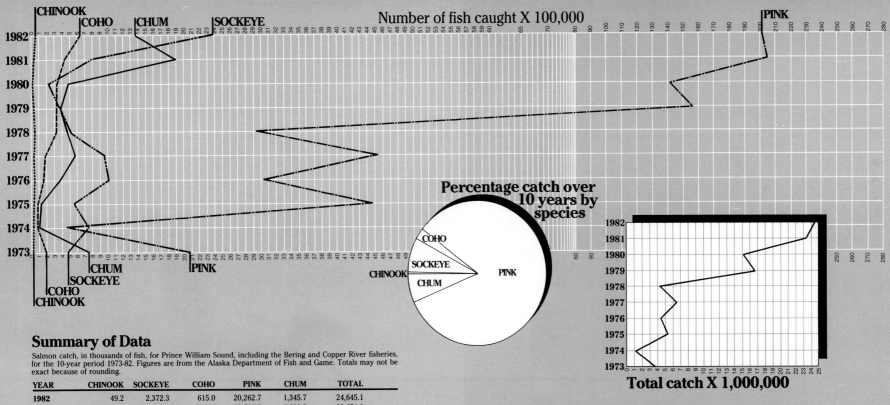

Number of fish caught X 100,000

CHINOOK COHO CHUM SOCKEYE PINK

1982 1981 1980 1979 1978 1977 1976 1975 1974 1973

CHUM SOCKEYE PINK
COHO
CHINOOK

Percentage catch over 10 years by species

CHINOOK COHO SOCKEYE CHUM PINK

Total catch X 1,000,000

Summary of Data

Salmon catch, in thousands of fish, for Prince William Sound, including the Bering and Copper River fisheries, for the 10-year period 1973-82. Figures are from the Alaska Department of Fish and Game. Totals may not be exact because of rounding.

YEAR	CHINOOK	SOCKEYE	COHO	PINK	CHUM	TOTAL
1982	49.2	2,372.3	615.0	20,262.7	1,345.7	24,645.1
1981	21.9	798.4	423.9	20,519.6	1,890.2	23,654.0
1980	8.6	208.7	337.1	14,161.0	482.2	15,197.7
1979	20.1	369.6	315.7	15,638.3	349.6	16,693.3
1978	30.4	505.5	312.9	2,917.5	489.8	4,256.1
1977	22.9	943.9	179.4	4,536.5	573.2	6,255.8
1976	32.8	1,009.1	160.5	3,022.4	370.7	4,595.4
1975	22.3	546.6	84.1	4,453.0	101.3	5,207.4
1974	20.6	741.3	76.0	458.6	89.2	1,385.8
1973	22.6	473.0	199.0	2,065.8	740.0	3,500.6
TOTAL	251.4	7,968.4	2,703.6	88,035.4	6,431.9	105,391.2
AVERAGE	25.1	796.8	270.3	8,803.5	643.2	10,539.1
% Each Species 10 Yrs.	.2%	7.5%	2.5%	83.5%	6.1%	
% Of Statewide Catch For Species, 10 Yrs.	3.7%	4.2%	10.3%	23.3%	8.8%	ALL: 15.7%

Limited entry permits held for Prince William Sound (September 1982) include: 270 purse seines, 543 drift gill net, and 30 gill net.

67

Cook Inlet — How It Was

Editor's note: This volume touches lightly on the rich history of Alaska's commercial salmon fisheries. That history deserves special attention in a separate presentation. We have included this brief history of Cook Inlet's salmon fishery as a sampler, and to give dimension to the present report.

One of the earliest records of commercial salmon transactions for Alaska dates to 1786, when two veterans of Capt. James Cook's expedition, Captains Portlock and Dixon, sailed directly to Cook Inlet from Hawaii on a trading expedition between northwest America and China, by way of the South Seas.

Russians that the two captains encountered at the mouth of Cook Inlet helped them find their way into Port Graham Bay, where the Russians' Aleut sea otter hunters traded the English fresh salmon for Hawaiian yams.

Alaska Packers, the Pacific Whaling Company, and the Northern Packing Company, all long gone from Alaska today, built salmon canneries in the Kenai-Kasilof areas in the late 1880s and early 1900s, and commercial exploitation of the inlet's salmon started. Within a few years salmon canneries and salteries were scattered from Port

Set-net fishermen commonly lay their net over the skiff and pull themselves from one end of the net to the other, picking fish and untangling the net as they go. This pair of fishermen are in Chinitna Bay, lower Cook Inlet.
(Jim Rearden, staff)

Chatham, at the tip of the Kenai Peninsula, to Anchorage, at the head of the inlet.

Salmon were for the taking in early years, with no restrictions on when, how, or where they were caught. Nets were strung in rivers or in salt water — where salmon were most easily caught. Earliest reliable catch records for sockeye salmon reveal that in 1893, 170,000 of these firm red-fleshed fish were canned in Cook Inlet. Like other salmon fisheries, Cook Inlet's catch increased rapidly: in 1900, 585,000 flopping reds were hauled from the inlet's cold waters and shoved into cans. This figure doubled by 1911, and by 1924, when first government controls appeared, the red salmon catch annually fluctuated from about 1 to 1.5 million of the silvery-scaled, rich, 6- to 8-pounders. Usually about 13 of the inlet's red salmon go into a standard case of 48 one-pound cans.

The first Cook Inlet commercial cannery fishermen used gill nets, staking and anchoring them along the shoreline where most of the fish travel. Fish swimming in the silty water cannot see, and blunder into the nets.

In 1885 the first salmon trap to be built in Alaska appeared in Cook Inlet, and in a short time both deep-water piling traps and hand traps ringed the inlet. Hand traps were constructed of slim spruce poles that could be carried and driven into Cook Inlet's blue clay bottom by hand.

All salmon traps worked on the same principle. Salmon follow the shoreline as they seek their home stream. A lead — simply a chicken wire fence attached to pilings or the hand trap poles — was constructed from shallow water perpendicular to the shoreline. Many ran hundreds of yards. At the end of the lead a "heart" was constructed, into which salmon could easily swim. When they tried to find their way out it was another matter. They would mill, eventually working into the next section of the trap, called the pot, or spiller, where they remained alive until brailed into a salmon tender — the vessel that hauled them to a cannery.

The first regulations for the inlet's salmon fishery appeared in 1924 — just nine lines of type. Signed by Herbert Hoover, Secretary of Commerce, the new regulations stopped commercial fishing for salmon after August 10; prohibited fishing within one statute mile of all salmon streams except for the Kenai and Kasilof rivers, where the closure was two miles; prevented fishing above a line from Point Possession to the western limit of the closed area around the mouth of the Susitna River; and prohibited fishing for salmon in Chenik Inlet, Kamishak Bay.

From these first simple rules the commercial regulations for salmon in Cook Inlet have evolved today's nearly 14 confusing pages of finely printed restrictions.

By 1930 a pattern had been set that was followed for nearly a quarter of a century. Fishery management headquarters for all of Alaska was in Seattle, where Bureau of Commercial Fishery personnel spent most of the year.

Annually in early May the 60-foot

A brailing net dumps more sockeye into the hold of a salmon tender. (Freda Shen)

wood-hulled slow-speed six-cylinder Washington diesel-powered USFS (for U.S. Fisheries Service) 1923-built *Teal* left Seattle and steamed through Southeastern Alaska delivering freight and passengers — mostly federal employees and their families — to various fishery stations, then she bucked across the stormy Gulf of Alaska to Seldovia, at the entrance to Kachemak Bay on the lower inlet.

The *Teal* remained part of Cook Inlet's salmon fishery for more than four decades, for she was transferred to the Alaska Department of Fish and Game at statehood, and she continued to patrol the inlet until the mid-1960s. She was finally sold at auction. When sold the old vessel was still sound, with the original power plant. Afterward she saw use as a tug in Southeastern Alaska. It was said she did an honest 7 knots — with or without a tow.

Seldovia was her headquarters for many years. Captain Roy L. Cole (Cap), who was skipper of the tough oak-framed vessel for most of the 1930s and 1940s, was also the fishery warden or fishery manager, as he was variously titled. U.S. commissioner's court from Kenai, Seldovia, and Anchorage, was often held in the varnished-wood saloon of the *Teal* when Cap had a case against some erring fisherman or salmon packer.

Fishermen registered their salmon nets with Cap aboard the *Teal* — one of the requirements for fishing during those years. Canneries radioed him their salmon pack reports each week. He poked the bow of the *Teal* into the

The boat harbor at Ninilchik.
(Francis Caldwell)

upper inlet, and into the shallow, rocky, and stormy waters of Kamishak Bay. On weekends when salmon traps were supposed to be rigged so they couldn't catch fish, Cap would snoop along the shoreline checking them.

In those years Alaska was a land with little public transportation, and roads were but a dream. Cap provided transportation for government surveyors, engineers, nurses, judges. The most famous passenger ever to tread the decks of the *Teal* was President Warren Harding, who made a brief trip on Cook

Inlet aboard the then-new vessel when he traveled to Alaska in the summer of 1923 to drive a golden spike symbolizing completion of the Alaska Railroad.

Cap Cole's logbooks also included the names of Clarence Rhode, Frank Dufresne, Gren Collins — all of the Alaska Game Commission; and L.J. Palmer, principal biologist, Bureau of Biological Survey; Fred Ball, U.S. Forest Service; District Judge Simon Hellenthal, and dozens of others.

From the *Teal* Cap Cole sent out high-speed patrol skiffs which could

venture into the shallow tidal flats of the upper inlet, and into the thin water on the west side of Kalgin Island. They were pushed by the back-breaking and cranky 22- and 33-horsepower outboard motors of the day. One, Cap's records reveal, could make 17 knots.

Cole had crews install salmon-counting weirs at English Bay Lakes, at Swanson River, Bishop Creek, Fish Creek, Packer's Creek (on Kalgin Island), and one at Chenik over in Kamishak Bay, where a trickle of a stream allows red salmon to almost crawl across a low pass to reach and spawn in Chenik Lake. Once, the story is told, Cap thought he could make it easier for reds to swim into Chenik Creek, so he blasted a barrier near the mouth. The experiment failed, for it made it more difficult for the fish; red salmon piled up there annually for years waiting for rain to raise the water level enough for them to swim past the barrier Cap had created.

Male chum salmon in spawning condition. Huge doglike teeth of such fish may be reason they are called "dog" salmon. (Matt Donohoe)

Cap Cole's annual reports, all "respectfully submitted, Roy L. Cole, Master, U.S.F.S. *Teal*," written in the dry governmentese of the time, are rich in the history of the great salmon runs of Cook Inlet.

At the end of each commercial salmon season Cap and his crews walked along the major spawning streams they could reach, counting spawning salmon.

In 1930, which is the earliest of Cap's annual reports now available within the state, 9 deep-water and 13 beach independent salmon traps were fished by "Tyonek Natives," Erick J. Fribrok, Simson Chickalusion, Knik Pete, Johnson and Johnson, Paul Shadura, and Julius Kallandar, among others.

Longtime Cook Inlet residents know these names — some descended from early Russians, others from the Native community, and still others were latter-day Scandinavians who tended to pioneer the North.

"Drift gill nets were more extensively used this year [1931] than heretofore. July 18, during a heavy storm, 19 gill-net power boats were anchored in the Kenai River for shelter," Cole reported.

In 1932, "Gill-net fishing opened May 25," and fishing was conducted largely in the shallows along the west shore from Harriet Point to the Susitna River. This was the early fishery for king salmon. That year the effects of the Great Depression must have taken its toll on the salmon industry, for Cole recorded but nine salmon canneries operating in the inlet, compared with 19 the previous year.

In 1934 Cap Cole ran the *Teal*, while William B. Berry was the warden. With a more-than-passing resemblance to an 1800s English Empire public servant, at the end of the season Berry wrote his "Report on Condition of Natives, Cook Inlet," saying, "The Natives . . . were in better financial condition than they have been for some time past." And, "The Natives will do quite a little trapping this winter and this will add substantially to their income."

Although the red salmon catch averaged well over 1 million fish a year at the time, in 1935 the Bureau of Commercial Fisheries and the salmon packers of the inlet decided that the Dolly Varden trout was destroying salmon by eating their eggs on the

spawning grounds. The packers contributed to a fund administered by the bureau from which a bounty of 2.5¢ each was paid for the tails of Dolly Varden.

At English Bay, where that year 15,851 red salmon were counted through the weir, 6,101 Dollies were trapped and destroyed. Dollies were also taken with nets and traps from Lake Tustumena and tributary streams of the lower Kenai River.

Warden Berry wrote, "There are no rainbow trout in any of the rivers or streams from the lower part of the Kenai River to Point Gore, this area being investigated very thoroughly."

It wasn't thorough enough: we know today that Anchor River, Deep Creek, Ninilchik River, and Stariski Creek are fine rainbow trout streams.

In 1936 "aeroplanes" were first mentioned in Cole's reports as being used to patrol the inlet salmon fishery, with 20 hours of flight contracted from Anchorage pilots.

Cap Cole took over as warden again in 1937, 1938, 1939. In 1939 a total of 27 hand traps and 25 pile traps fished Cook Inlet, with 14 beach seines and 20,805 fathoms of red salmon gill net. Fourteen drift gill nets were used for fishing for red salmon. And Cole reported that 32,547 salmon-fry-and-egg-eating Dolly Varden were killed. An airplane from Star Air Service (destined to eventually become today's Alaska Airlines) was chartered for one patrol along the east side of Cook Inlet — Anchorage to Kenai — thence to Kustatan, and return to Anchorage via

NcNeil River brown bear, lower Cook Inlet, and a leaping chum salmon. (Tom Walker)

Sockeye salmon entangled in a torn gill net, being brought aboard a drift gill-net boat. (Freda Shen)

the west side of the inlet. "Two hours flying time were consumed, covering approximately 90 nautical miles," Cole reported.

He reported that airplanes were not really suitable for salmon fishery patrols, for they couldn't land whenever they wanted because of the rough water. Some violations had to be overlooked.

Under "Conditions of Local Whites and Natives" (Cap added "Whites" to the earlier format) Cole reported, "No labor troubles of any consequence. Both Natives and Whites were employed in the salmon fishery to the fullest extent."

A fisherman using company-owned gear that year of 1939 received from the cannery 14¢ for each red salmon, 70¢ for each king, regardless of size: average fish weighed 20 pounds, others went up to 80 pounds or more, while chum and pink salmon brought the incredible price of 4¢ each. "Medium reds" (coho salmon) brought 14¢ each. Independent fishermen who owned their nets and traps received about one-third more.

In the season of 1945 the 55 salmon traps of the inlet caught 1.6 million salmon of all species, seines caught 1.1 million, and gill nets caught another 800,000. The best trap catch was by the Fidalgo Island Packing Company trap number 6, license number 45-040, at Flat Island, at the tip of the Kenai Peninsula and the entrance to the inlet. It caught 149,360 salmon — 3,285 sockeye, 195 kings, 135,780 pinks, 8,540 chums, and 1,560 coho.

It was a landmark year for set-net and trap fishermen, for that was the year that Judge Anthony Dimond made his famous "he who fishes first" ruling for stationary salmon gear, the guidelines of which are presumably still followed by fishermen and Alaska's courts.

The late Luke Fisher, tall, slim, story-tellin' and easy going, left his Seldovia home early that year and headed for Granite Point for his set net sites. On April 22 he drove the stakes for his first gill net, marking where he intended to fish for the coming season.

Selter Hale, fishing for a Mrs. Everett, who owned the sites, also at Granite Point, drove stakes on May 11. One set of stakes was closer than the law allowed to those driven by Fisher.

When fishing started Luke got his set net into the water first; Hale strung his too-close net next. Cap Cole, on June 1, finding two set nets fishing closer than the law allowed, cited both into court. Judge Dimond found in favor of Fisher, ruling, "He who first occupies a certain location or site for purpose of fishing that site with any particular type of fixed gear is entitled to it because he is in possession of it."

He pointed out that set gill-net fishing is virtually a new enterprise each season, for which sites must be located anew and stakes driven. He wrote, "He who first, in good faith, and with proper and adequate means begins the occupation of such a site, will be protected in his priority, pro-vided he diligently brings his work to completion as actually to fish the site,

Kenai Packers drift boats tied up at the Kenai dock.
(Stephen Bingham)

and will not lose priority to someone who commences to occupy a site within prohibited distance at a later date."

War-time inflation was reflected in the prices that Luke Fisher, Hale, and others received for salmon in 1945. Independent fishermen south of Boulder Point received 28¢ each for sockeye, 1.13⅓¢ for chinook, 8⅔¢ each for pinks and chums, and 28¢ for coho.

By 1947 salmon fishermen were restricted to fishing only five and a half days a week for the May 20 through August 8 season. In 1950 this became five days a week, with three 24-hour closures during the season in addition to the regular weekend closures.

By 1955 fishing was allowed only two days a week from May 25 to July 27 in order to protect chinook and sockeye, and five days a week July 28 through August 4.

About 1950 the fisheries manage-ment office was moved from Seattle to Anchorage, where agents remained year-round. Absentee management was finished. In the early 1940s Cap Cole ceased making the long journey north each year with the *Teal*, and others took over management of Cook Inlet's fishery. But Cap wasn't for-gotten. In 1953 owners of six trap-owning canneries recommended to the U.S. Fish and Wildlife Service (the agency responsibile for Alaska's

Set-net fishermen mending gill nets at Clam Gulch, Cook Inlet. *(Sharon Paul, staff)*

fisheries in those years) that drift gill nets be restricted to the extreme southern part of the inlet (south of 60 degrees north latitude), and be shortened from 150 to 100 fathoms — a typical move in those years. Canneries which depended upon drift gill nets responded with their own recommendations.

In their pitch to the USF&WS the trap-owning canneries wrote, "The history of the successful salmon runs in Cook Inlet has been without parallel anywhere else in Alaska, on the Pacific Coast, or in any other known salmon producing regions of the world." (They weren't going to miss any bets.)

They went on, "This outstanding example of management and conservation was not happenstance, but was largely the result of the work of a former employee of the Bureau of Fisheries who always gave first consideration to principles of conservation in all decisions affecting management and regulation of this fishery throughout the many years the district was under his immediate supervision."

George Black, Tom Costello, Jack Skerry, all college-trained fisheries scientists, took turns managing Cook Inlet's fisheries. Twin-engined Grumman Goose amphibious airplanes were used during the salmon season to patrol the inlet from the Anchorage base.

The last year that the federal government managed the Cook Inlet (and all of Alaska's) fishery was 1959. In that year inlet fishermen caught the fewest salmon taken since the earliest days of the fishery — only 1.3 million, about one-third the average of the previous five years' annual catch.

It was now the state's job to rebuild the salmon fishery.

Cook Inlet — How It Is Today

Cook Inlet is a tapered bay that extends north and east for about 220 miles from the northern Gulf of Alaska. During the 10 years 1973-82, Cook Inlet's commercial salmon fishermen caught an annual average of 4.6 million salmon, or about 6 % of the statewide catch.

Cook Inlet's waters, in common with the other great gill-net fishing districts of Alaska, are silty. Most major streams that pour into the inlet carry silt, much of it from glaciers. The biggest include the great Susitna River, at the head of the inlet, the nearby Little Susitna and Matanuska rivers, and, about halfway down the inlet on the east side, the Kenai and Kasilof rivers.

The northern 190 miles of the inlet are perpetually murky, and the line between clear and murky water is often abrupt. Usually this line is found at about the latitude of Ninilchik, but it moves north with large tides, and south with minus low tides. Above the Forelands — which is the sharp narrowing about two-thirds the way up the inlet — the inlet appears to be liquid mud.

The inlet is rimmed on the west by volcanic peaks: cone-shaped Mount Augustine, which last errupted in 1975, is an island in the lower inlet; Mounts Iliamna and Redoubt are perpetually snow-capped steam-venting peaks of around 10,000 feet in the Aleutian Range; and Mount Spurr, which last errupted in the early 1950s, is an 11,020-foot peak in the Alaska Range. The generally low shoreline of the Kenai Peninsula lies on the eastern side of the inlet.

The climate is primarily northern maritime, although colder continental conditions prevail in the northern interior basins of the inlet. Cool cloudy summers, wet autumns, variable but typically cold winters, and relatively cool, dry springs are found.

About half of Alaska's human population lives in the Cook Inlet basin. Anchorage (pop. 173,000), Alaska's largest city, lies at the head of the inlet. Kenai (4,300), and Soldotna (2,320) are found on the upper Kenai Peninsula. Homer (2,211) and Seldovia (473) lie on opposite sides of Kachemak Bay, a 30-mile-long, deep, clear-water bay which juts easterly from Cook Inlet near the tip of the Kenai Peninsula. Ninilchik (336), Tyonek (239), English Bay (53), and Port Graham (162) are villages on the shores of the inlet.

Cook Inlet has long been known for its stable salmon fishery: cycles of abundance and scarcity are not generally as severe as those of many other of Alaska's salmon fisheries. The great diversity of spawning streams and lakes that surround the inlet, with

the broad range of climatic conditions, probably account for this: when severe weather causes mortality in streams and lakes of one part of the inlet, moderate conditions may be found elsewhere, with consequent higher salmon survival.

The inlet has the largest tides in Alaska, with maximum high tides reaching nearly 33 feet at the northern end. The inlet narrows to the north, compressing the incoming tide. At narrow points, as between the Forelands, and between Harriet Point and Kalgin Island, tides may flow at 8 knots or more. Tidal swirls and rips are common throughout the inlet, making small boat navigation interesting.

Fishermen pay a penalty for fishing in the gill net districts of Cook Inlet, for compared with many other fishing

A bear-killed red salmon. Alaska's coastal bears feast on spawning salmon. (Tom Walker)

Skipper Pete Islieb picking sockeye salmon from his drift gill net. Nylon fish gloves hold slippery fish. (Freda Shen)

districts, the inlet is rough and difficult to navigate. Distances are considerable — with about 70 miles of open water between the northern and southern boundaries of the drift gill net (Central) district and 30 to 40 miles between the eastern and western shores. The usually rough seas that build with strong southwest summer afternoon breezes whooping up the inlet toss the average 30- to 35-foot drift gill-netters about violently, often forcing a stop to fishing. The 20-foot set-net fishermen's skiffs are difficult to launch, and the nets hard to pick.

Although the upper inlet is not considered important as a salmon feeding and rearing area, it is the route for the five species of salmon that leave their home streams and go to sea; they return as adults through the swift tides and murky turbulent waters to their home streams.

Sockeye salmon, with its ruby red flesh, was the lure that brought early salmon packers and fishermen to the inlet. It is still the main lure. The fish arrive in a great flood in July, pouring from the Gulf of Alaska through Shelikof Strait and the wind-and-tide-tossed waters between the Barren Islands and the tip of the Kenai Peninsula. Scientists who have tagged these fish in the lower inlet found that most arrived on the west side of the lower inlet to swim northeast, heading mostly for the great Kenai and Kasilof rivers. The Kenai River is the largest producer of sockeye salmon in the inlet, and the huge Susitna is second largest. Depending upon weather, Susitna-bound fish may follow the east or west shore of the inlet as they seek the mouth of the Susitna. Others home on numerous other streams around the inlet.

The multiplicity of streams connected to lakes that flow into all sides of Cook Inlet (sockeye generally spawn only in drainages where there are lakes) has resulted in a wide variety of races or types of sockeye salmon: some are big, deep, and heavy. Others are long and slim. A few are tiny, barely 2 or 4 pounds.

Cook Inlet's sockeye salmon catch, on average, is the second largest for that species in Alaska, coming only after that of Bristol Bay, although the inlet produced only about 8.5% of the sockeye caught in Alaska during the years 1973-82. Normally about 60% of returning Cook Inlet sockeye are five-year-olds; from 15% to 20% are four-year-olds; and less than 20% are six-year-olds.

Sockeye is the money fish in Cook Inlet, and during 1973-82 an annual average of 1.6 million sockeye were caught — about 8.5% of the statewide catch of sockeye for the period. Sockeye made up 34.5% of the total inlet catch. Major runs of sockeye swim into the Kenai, Kasilof, Susitna, and Crescent rivers — all in the Northern and Central districts. The peak of the sockeye fishery is usually between July 15 and 20.

Salmon traps caught most of the inlet's sockeye for many years. Since statehood, and the banning of salmon traps, gill nets have caught virtually all

A fresh-caught pink salmon, its scales as shiny as new dimes, lays on the deck of the seiner that caught it.
(Jim Rearden, staff)

of the inlet's sockeye. By the early 1950s between 500 and 600 drift gill nets annually fished the silty waters of the inlet for reds — about the same number as today. However, fishing time allowed in those years was five 24-hour fishing periods a week: today the basic time is two 12-hour fishing periods a week.

From its start in 1893 the inlet's sockeye catch has reached or exceeded 2 million 10 times. Largest catch ever, of 3.1 million, was made in 1982.

While sockeye was the original attraction for packers and fishermen, in recent years the other four species have also become important.

Pink salmon are the most abundant salmon in Cook Inlet. For the years 1973-82 this species made up 39.6% of the total catch (numbers of fish), with an annual average catch of 1.8 million. This was about 4.4% of the statewide catch of pinks of those years.

Pink salmon in the gill net districts of the inlet (north of Anchor Point) have a distinct even-year cycle: catches range to perhaps 100,000 fish in odd years

(although there was an inexplicable catch of 554,000 pinks in these districts in 1977 — the largest ever odd-year pink catch) but it ranges from 1 to 2 million in even years.

Major pink-producing streams include Kenai River and the Susitna River at the head of the inlet: the clear-water Talachulitna River, a tributary of the Susitna, is probably the most important pink producer, with as many as 1 million pink salmon spawners in some years.

Prior to about 1940 the smaller (average 3 to 4 pounds), spotted, and soft-fleshed pink salmon were incidentally caught by fishermen using large-mesh (5¼-inch or larger) gill nets seeking sockeye, chum, coho, and chinook salmon, but now pinks are of importance on their own, and pink gill nets with a 4-inch or slightly larger mesh are commonly fished after the bulk of the sockeye have passed.

Pink salmon are the primary species caught by Cook Inlet's hand purse seine fishermen who fish south of Anchor Point. During the period 1948-61 the annual average catch here was around 480,000 pinks. Then, unexpectedly, in 1962, the catch soared to 2 million pinks. Two years later, in 1964, the catch was still high at 1 million.

From 1965 through the early 1970s the annual pink catch south of Anchor Point averaged 374,000. Catches during the early 1970s were poor. Then in the late 1970s a major change took place: the dominant even-year pink runs changed so that the odd-year pink runs became dominant. In 1979 the catch was 2.9 million, of which 375,000 were produced by a state hatchery in Tutka Lagoon — on the south side of Kachemak Bay.

The 1980 catch of 728,000 was well above the average. Then in 1981, the catch was 2.8 million pinks, of which 1 million were produced by the Tutka Lagoon hatchery. Instead of a 1% to perhaps 3% return of fry, biologists estimated that the return to the hatchery was perhaps 15% to 18% — an almost unheard-of survival rate for pink salmon. Credit was given to short-time rearing and feeding of the fry before they were released in Tutka Lagoon to make their way on to their 14-month sea-faring odyssey.

In 1982 the Tutka Lagoon hatchery return was only 231,800 pinks — far below that expected.

Chum salmon in upper Cook Inlet are a "me too" fish: they are caught

Susan Beeman hanging a set gill net (attaching lead and cork lines). She's working on the cork line. (Tom Walker)

mostly by gill-net fishermen who are seeking the more valuable sockeye. They are about the same size as the sockeye and nets designed for sockeye are equally effective in taking chums. Chums made up about 19% of the Cook Inlet catch for the 10 years 1973-82, with an average annual catch for those years of about 871,000. Normally about 85% of these are caught by gill-net fishermen north of Anchor Point, and the balance by hand purse seiners fishing in the sheltered bay of the lower Kenai Peninsula, and in the often-stormy Kamishak Bay.

Chums are predominantly four-year fish in Cook Inlet, and they tend to have an even-year cycle.

For the gill-net fishery, the peak of the chum run occurs about a week after that of the sockeye — or sometime shortly after about July 20.

The vast Susitna basin produces about 90% of the chums of the gill-net fishery of the inlet. Chinitna Bay and Kamishak Bay, both on the west side of the lower inlet produce chums, as do certain bays of the lower Kenai Peninsula — Rocky, Windy, Port Dick, Seldovia, and Port Graham bays, among others.

Peak catches of chums in Cook Inlet occurred in 1964 (1.4 million), 1977 (1.3 million), and 1982 (1.5 million).

Coho salmon are another "me too" salmon in Cook Inlet that are caught mostly by fishermen who set their nets for the more valuable sockeye. The average annual coho catch for the inlet, 1973-82, was 30,400, or about 6.5% of the inlet's salmon catch. It was also

Paul Jones, of Homer, wrestles with his drift gill net, readying it for rolling it onto his gill-net reel.
(Jim Rearden, staff)

This 78½-pound king salmon was caught on sport tackle in the Kenai River by the late J.J. Smith, who stands behind the huge fish. Smith is 5' 6" tall. (Jim Rearden, staff)

11% of the total statewide coho salmon catch for that 10-year period, which says that Cook Inlet is an important coho producer.

At least 95% of the annual commercial coho catch is made by gill-net fishermen north of Anchor Point. Largely a four-year fish, the inlet coho tends to have an even-year cycle, but the cycle is not as pronounced as that of the pink salmon.

Coho salmon are found in almost every spawning system of the inlet, and they tend to be rather evenly distributed around the inlet — more so than any other species. They start arriving in mid July, and some coho are still arriving as late as October. The peak of the coho run for the upper inlet is in late July.

Because the coho arrives and enters streams in the fall when water tends to be high and murky from fall rains, fishery managers seldom have much information on the numbers that reach spawning grounds.

The smallest coho salmon catch in 50 years occurred in 1972 when only 83,000 were taken. Remarkably, the coho has made a tremendous recovery: the largest catch of coho ever for Cook Inlet was in 1982, when 765,400 of the silver-sided salmon were landed.

King salmon, silvery, black-spotted, huge, rich-fleshed, are now relatively unimportant to Cook Inlet's commercial fishermen. For the 10 years 1973-82, an annual average of 12,470 were caught — less than 1% of the inlet's salmon catch, and 1.8% of the statewide commercial catch of kings.

It was not always so. Greed almost destroyed Cook Inlet's kings. From 1924-40, commercial fishermen caught an average of 66,000 Cook Inlet kings each season. Then the catch increased: from 1941 to 1953, the catch averaged 109,000, with the largest catch ever in 1951 when 187,000 king salmon were caught.

After that it was all downhill, until shortly after statehood when a total closure for commercial and sport fishing for king salmon throughout the Cook Inlet basin allowed stocks to rebuild.

By 1977 more than 100,000 spawning king salmon were counted in the Susitna basin, and the species now seems secure.

Virtually all the king salmon caught by commercial fishermen in the inlet are taken by gill-netters north of Anchor Point. Kings enter the inlet in two separate runs: the early run is the big one, and it is bound for the Susitna basin. The bright strong sea-fresh fish arrive in the inlet about May 25, and by mid-June most have reached the silty Susitna and scattered to the many clear and not-so-clear streams where they spawn.

The second run, which arrives in June and continues through most of July, is bound mostly for the Kenai-Kasilof rivers. This is the run from which most of the commercial harvest is taken today. The current late June opening of commercial salmon season in the inlet comes well after the Susitna-bound kings have reached their spawning areas.

Salmon catch for
Cook Inlet

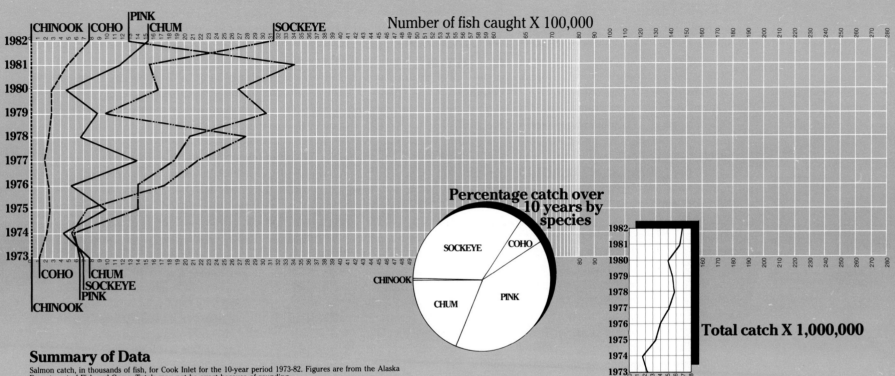

Number of fish caught X 100,000

Percentage catch over 10 years by species

Total catch X 1,000,000

Summary of Data

Salmon catch, in thousands of fish, for Cook Inlet for the 10-year period 1973-82. Figures are from the Alaska Department of Fish and Game. Totals may not be exact because of rounding.

YEAR	CHINOOK	SOCKEYE	COHO	PINK	CHUM	TOTAL
1982	1.9	3,147.6	765.4	1,278.1	1,524.2	6,735.2
1981	13.3	1,549.5	495.9	3,406.4	1,169.6	6,634.7
1980	14.2	1,643.0	285.9	2,676.1	463.2	5,082.5
1979	15.0	988.8	287.8	3,049.2	873.4	5,214.1
1978	19.0	2,778.1	225.3	2,041.7	645.5	5,709.6
1977	15.0	2,154.1	195.8	1,846.3	1,379.5	5,590.8
1976	11.3	1,722.3	211.9	1,393.2	520.6	3,859.4
1975	4.9	713.0	233.6	1,399.8	973.4	3,324.7
1974	6.8	524.6	206.6	534.3	416.1	1,688.4
1973	5.3	699.2	106.5	633.6	783.1	2,227.8
TOTAL	124.7	15,920.2	3,014.7	18,258.7	8,748.6	46,067.2
AVERAGE	12.4	1,592.0	301.5	1,825.8	874.8	4,606.7
% Each Species 10 Yrs.	.2%	34.5%	6.5%	39.6%	18.9%	
% Of Statewide Catch For Species, 10 Yrs.	1.8%	8.5%	11.5%	4.4%	12%	ALL: 6.8%

Limited entry permits held for Cook Inlet (September 1981) included 590 drift gill net, 748 set gill net, and 84 hand purse seine.

83

The Kodiak District

Kodiak Island, in the turbulent western Gulf of Alaska, 100 miles long, 60 miles wide, is Alaska's largest island. Thirty-mile-wide tide-and-storm-tossed Shelikof Strait separates lovely Kodiak Island from the Alaska Peninsula.

A spine of steep, rugged mountains runs the length of Kodiak; deep fjords and bays indent the coastline into which hundreds of short streams and rivers flow. Many small lakes and ponds are scattered across the glaciated surface of lush Kodiak and nearby spruce-covered Afognak Island. Only the northern end of Kodiak Island has spruce trees: the rest is grass, alder, and a wide variety of brush and forbes. The Karluk and Red rivers, both about 25 miles long, drain much of southwestern Kodiak Island. Karluk River, including 12-mile-long Karluk Lake, and Dog Salmon River, including 9-mile-long Frazer Lake, are among the most important salmon producing systems.

The annual average 60 inches of precipitation and the cool-to-mild maritime climate produce luxurious near-tropical plant growth. Severe storms with high winds are common. The temperate North Pacific waters that surround Kodiak are alive with a wide variety of marine life.

The Kodiak area is fished by purse seines, beach seines, and set nets. During the 10 years 1973-82 Kodiak's salmon fishermen caught an annual average of 10 million salmon of all five species — or about 15% of the total statewide catch of salmon. Most of these fish were pink salmon (85%), for the many short streams of Kodiak are ideal for this species. Chum salmon are the second most abundant salmon of the Kodiak area, and in recent years about 7.2% of the catch has been of chums.

The town of Kodiak (pop. 4,746) on the north end of Kodiak Island is the main human and trade center. Akhiok (105), Port Lions (215), Ouzinkie (173), Karluk (96), Old Harbor (334) and Larsen Bay (144), are villages on and around Kodiak Island. The Alaska Marine Highway (state ferry), and regular airline service provide transportation to and from the island. Few roads exist, and most local transportation is by small plane or boat.

Purse, hand purse, and beach seines are used to catch salmon in all Kodiak districts except for Olga and Moser bays, where only set gill nets are permitted. Set nets also fish in a few other areas on the west side of the island.

King salmon catches are minor in the Kodiak area, averaging about 1,100 fish a year during the 10 years 1973-82. The Karluk and Red rivers have the only natural king salmon runs. Kings were apparently successfully introduced into the Dog Salmon-Frazer Lake system in the mid-1960s.

Red salmon dominated the Kodiak fishery in the late 1890s and early 1900s, but today it is only the third most abundant species. Average annual sockeye catch 1973-82 was 683,500, making up 6.8% of the Kodiak catch. There are more than 30 sockeye-producing systems in the Kodiak fishing area, but only four of these —Karluk, Red River, Upper Station, and Frazer Lake are of real importance.

Frazer Lake, which drains into Olga Bay, is one of Kodiak's largest lakes. It had no salmon runs because of an impassable (to salmon) falls. In 1951 the Territorial Department of Fisheries introduced red salmon into the lake, and followed up the egg plant with introduction of live spawners over a period of several years. As sockeye returned to the Frazer they were netted and carried over the falls in backpack tanks. The run gradually increased. In 1962 fishways were built which put an end to backpacking salmon: they can now swim across the falls on their own. This run, one of the few successful man-established sockeye runs anywhere, is still increasing: in 1982, 400,000 eager, flipping and splashing sockeye swam through the fishway and into Frazer lake to spawn.

Karluk Lake, on the west side of Kodiak Island, once sustained the greatest red salmon fishery known for any single river and lake anywhere. The 12-mile-long lake and 24-mile-long river which flows through Karluk Lagoon and into Shelikof Strait, produced a catch of 4 million in 1901. Between 1887 and 1928, the average annual catch of sockeye from the Karluk was 1.9 million.

Salmon and crab fishing boats crowd the Kodiak harbor. Some boats are used for both, and rigging is changed seasonally. (Sharon Paul, staff)

Pink salmon in Portage Creek, Afognak Island. (Dennis Gretsch)

Since 1938, when 1 million were caught, the catch has declined. In 1982, although the escapement goal was 800,000 for the Karluk, only 132,000 red salmon arrived.

Despite years of research and good numbers of spawners over the years the Karluk red run hasn't increased: no one knows why. A large even-year run of pink salmon to the Karluk complicates the picture: how do you harvest large numbers of pinks without taking more sockeye than you want?

Each year, when regulations permit, Kodiak seiners run to Cape Igvak on the mainland, where they net good numbers of sockeye. A tagging study of these fish showed that most were bound for Chignik, to the south, and some were bound for Cook Inlet, to the north. If a weak run is expected at Chignik this fishery is curtailed.

The Alitak district, on the south end of the island, has historically been second only to the Karluk in producing large red salmon catches. Upper Station, Frazer, and Akalura are the major producers there.

The Kodiak red salmon catch for 1982 of 1.2 million was the second largest red salmon harvest in 34 years. The catch of 215,000 sockeye at Cape Igvak plus fair to good runs to Kodiak Island systems accounted for this.

Pink salmon eggs hatch and eventually the fry wriggle out of the gravel to almost immediately go to salt water. Stable water levels, clean gravel, cool summer and mild winter temperatures are major requirements for a pink salmon stream.

Most of the approximately 300 streams in the Kodiak management area meet these requirements, and most of them produce pinks. The bulk of Kodiak's pink salmon, however, are found in about 31 of the major rivers.

Today more than 90% of Kodiak's pink salmon are caught by purse seiners. Pink salmon were relatively unimportant in the Kodiak area until after about 1912. Catches from 1934 through about 1947 were especially good, averaging 8.5 million fish.

After a decline, catches again increased with the average annual catch from 1962 to 1970 being about 8.6 million.

The average annual catch for the 10-year period 1973-82 was 8.5 million. This was 22.5% of the statewide catch of pink salmon. Peak year was 1980 when 17.2 million were caught: other top years were 1937 with a catch of 16.7 million, 1962 with a catch of 14.1 million, and 1978, with 15 million.

In the past the odd-year cycle of pinks dominated. In the mid-1970s the even-year cycle was dominant, with weak runs in odd years. Since then both even- and odd-year catches have been good, averaging more than 12 million a year since 1976.

Important pink producers of Kodiak include Terror, Uganik, and Uyak rivers, on the west side of the island. The Karluk (of red salmon fame), Sturgeon and Red River districts on the west and southern end of the island are great pink producers — catches of 4 million or more in a season are not unusual.

A typical modern salmon seine type vessel (left) in the Kodiak harbor. Such boats are commonly also used for fishing for crab, shrimp, and herring. (Staff)

The Alitak district, on southern Kodiak, generally produces catches ranging from half a million to 2 million.

The east side of Kodiak Island, which faces the Gulf of Alaska, has many small pink salmon-producing streams and a few major rivers. There seiners commonly round-haul and brail board 2 to 4 million or more silvery splashing pinks during a season.

Afognak Island, just north of and across the channel from Kodiak Island, has a strong pink run, with yearly catches ranging from 50,000 to 2 million or more. A fishway built by the state on Perenosa River has made that system one of the best pink salmon producers on Afognak Island.

Chum salmon are second to pinks in abundance on Kodiak Island. For the 10-year period 1973-82, the annual average catch of chums in the Kodiak management district was 732,400 — which was only 7.2% of the total

Kodiak catch. It was also 10% of the total statewide catch of chums for that period.

Like most other salmon fisheries dominated by one species — in this case pink salmon — management decisions are based on when and where it is proper to catch pinks. If chums are present, they are also caught.

Kodiak chums generally use the same spawning systems as pinks. One slight difference between the two species is the different time of arrival in bays and estuaries: pinks arrive around July 1, chums around July 15. Both are present in salt water until about September 1.

Chum catches appear on the increase in recent years: more than 1 million chums were annually caught each year 1980-82. Previously the only recent catches of 1 million or more were in 1960, 1971, and 1972. The 1982 chum catch of 1.2 million ranked as Kodiak's third largest (behind 1971 and 1981).

Coho salmon catches at Kodiak are generally small. For the 10-year period 1973-82 the annual average catch was only 88,610. This figure is somewhat distorted by the spectacular and unexpected catch of 343,000 in 1982 — the largest ever coho catch at Kodiak. For the nine years prior to 1982 the average annual catch of cohos for Kodiak was 53,600.

Red salmon struggle to overcome Fraser Falls to reach their spawning grounds in western Kodiak.
(Gerry Atwell, USF&WS, reprinted from ALASKA GEOGRAPHIC®)

Salmon catch for
Kodiak Commercial Fishing District

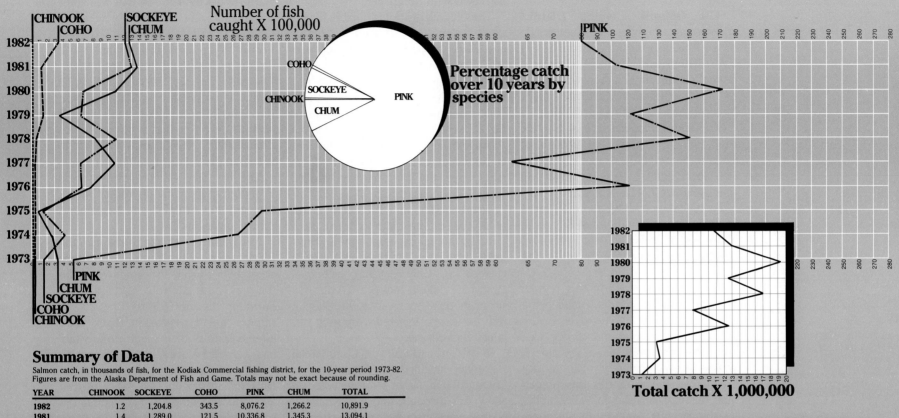

Number of fish caught X 100,000

CHINOOK COHO SOCKEYE CHUM PINK

1982 1981 1980 1979 1978 1977 1976 1975 1974 1973

PINK CHUM SOCKEYE COHO CHINOOK

COHO
SOCKEYE
CHINOOK
CHUM
PINK

Percentage catch over 10 years by species

Total catch X 1,000,000

Summary of Data

Salmon catch, in thousands of fish, for the Kodiak Commercial fishing district, for the 10-year period 1973-82. Figures are from the Alaska Department of Fish and Game. Totals may not be exact because of rounding.

YEAR	CHINOOK	SOCKEYE	COHO	PINK	CHUM	TOTAL
1982	1.2	1,204.8	343.5	8,076.2	1,266.2	10,891.9
1981	1.4	1,289.0	121.5	10,336.8	1,345.3	13,094.1
1980	.5	651.4	139.2	17,290.6	1,075.6	19,157.2
1979	1.9	630.8	140.6	11,285.8	358.3	12,417.4
1978	3.2	1,071.8	48.8	15,004.1	814.3	16,942.2
1977	.6	623.5	27.9	6,252.4	1,072.3	7,976.7
1976	.8	641.5	23.7	11,078.0	740.5	12,484.5
1975	.1	136.4	23.7	2,942.8	84.4	3,187.4
1974	.5	418.8	13.6	2,647.3	249.3	3,329.4
1973	.8	167.3	3.6	511.7	317.9	1,001.3
TOTAL	11.0	6,835.3	886.1	85,425.6	7,324.1	100,482.1
AVERAGE	1.1	683.5	88.6	8,542.5	732.4	10,048.2
% Each Species 10 Yrs.	.01%	6.8%	.8%	85.0%	7.2%	
% Of Statewide Catch For Species, 10 Yrs.	.16%	3.6%	3.3%	22.7%	10.0%	ALL: 15.0%

Limited entry permits held for Kodiak (September 1982) included 386 purse seine, 34 beach seine, and 187 gill net.

The Chignik District

The Chignik Fishing District spans about 175 miles on the south side of the Alaska Peninsula — from Imuya Bay to Kupreanof Point. A variety of coastal streams in this area produce all five species of salmon which are caught by local fishermen using purse seines.

The heart of the district is at Chignik itself, which includes the villages of Chignik (pop. 179), Chignik Lagoon (48), and Chignik Lake (138), about 160 miles southwest of Kodiak Island.

Perryville (pop. 108), about 50 miles south of Chignik, and Ivanof Bay (pop. about 30), 30 miles west of Perryville, are the only other villages within the district.

There are no roads in this region: most transportation is by small plane and boat.

Chignik is beyond the spruce tree line, and the largest plants are alders, which grow profusely. Hummocky grass-covered tundra fills this rugged land, and in most years snow lies on the nearby treeless peaks until mid-June. Fog and clouds are more common than sunshine. Brown bears roam the beaches and uplands, and in summer and fall, eagles and gulls quarrel over pink, chum, and coho salmon that run into the many short streams, as well as the red salmon which run into the deep, clear Chignik River, which drains 8-mile-long Chignik Lake, which in turn drains 6.5-mile-long Black Lake.

Both lakes lie across the low back of the Alaska Peninsula. It is but a short walk from the upper end of Black Lake to north-draining streams which flow to 20-mile-distant Bering Sea.

The Chignik district is notable for the great sockeye salmon runs to Chignik and Black lakes. During the 10 years 1973-82 Chignik fishermen lifted aboard an annual average of 2.1 million flipping, dripping salmon — 54.8% of which were sockeyes bound for those two lakes. Like Cook Inlet, the Chignik sockeye fishery has held up well under commercial fishing — since 1888, in fact, when the first pack of 13,000 Chignik reds was put up.

Between 1900 and 1924 an average of 16 pile-driven salmon traps, with virtually no management or enforcement of conservation laws, took annual catches at Chignik that averaged almost 1.3 million sockeye. To this day scientists don't know why the Chignik red run wasn't depleted from such sustained high catches.

In 1922 the Bureau of Commercial Fisheries installed a fish weir in Chignik River. Gates were installed to allow fish through, and white-painted boards were placed on the bottom so fish could be counted as they swam across. The weir was installed annually, except for the war years, and when water was too high, until after statehood, providing an accurate count of the number of spawners reaching the two big lakes.

In 1955 the salmon industry-financed Fisheries Research Institute of the University of Washington started a long-term research program on Chignik red salmon. The Alaska Department of Fish and Game later joined this work.

The researchers made a basic and startling discovery, one that has since drastically altered management of Chignik's reds. They learned that instead of one great run of salmon, two distinct major runs enter Chignik River. The early run, which arrives mostly in June, swims steadily up Chignik River, through Chignik Lake and through shallow, winding Black River, which drains Black Lake, and goes on to spawn in Black Lake tributaries.

The next fish — the bright, shiny and lively red salmon that pour into Chignik River in July, August and September, spawn in Chignik Lake.

It is relatively simple to manage the Black Lake and Chignik Lake runs separately so that fishermen can take the maximum allowable catch without harming the runs. Most Black Lake salmon spend a year in Black Lake after hatching, then spend three years in salt water before returning to spawn.

Chignik Lake salmon are different: they mostly spend two years in that lake after hatching, then descend to the sea, where they spend another three years before returning to spawn.

The old system of management — fishing heavily on the first part of the run, and curtailing the fishing later — resulted in reducing Black Lake fish by two-thirds between the years 1927 and 1939. At the same time, Chignik Lake salmon decreased by less than half. The startling but simple discovery of two runs of fish, each with its own time of return, and its own lake for spawning, has made the reasons clear.

Nowadays red salmon are still care-

Salmon catch for
Chignik Fisheries District

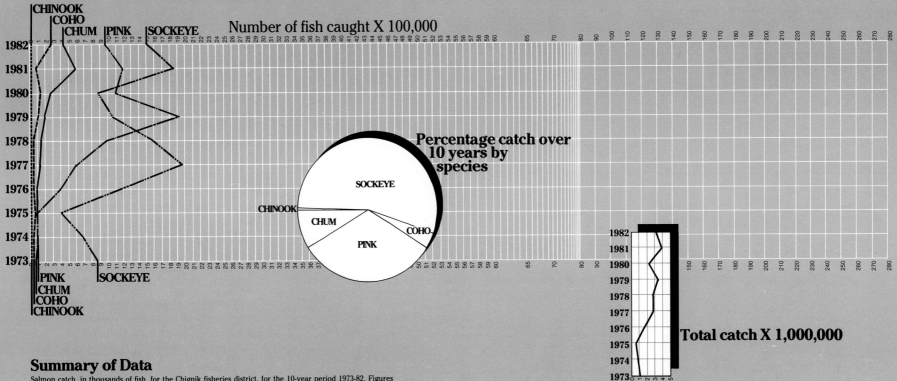

Number of fish caught X 100,000

Percentage catch over 10 years by species

Total catch X 1,000,000

Summary of Data

Salmon catch, in thousands of fish, for the Chignik fisheries district, for the 10-year period 1973-82. Figures are from the Alaska Department of Fish and Game. Totals may not be exact because of rounding.

YEAR	CHINOOK	SOCKEYE	COHO	PINK	CHUM	TOTAL
1982	5.2	1,509.2	289.2	942.7	403.5	3,149.8
1981	2.7	1,839.5	78.8	1,162.6	580.3	3,663.9
1980	2.3	860.0	119.6	1,093.2	252.5	2,327.6
1979	1.3	1,049.7	99.1	1,905.2	188.9	3,244.2
1978	1.6	1,576.3	20.2	985.1	120.9	2,704.1
1977	.7	1,972.2	17.4	604.8	110.5	2,705.6
1976	2.3	1,163.7	35.2	395.3	81.4	1,677.9
1975	.5	399.6	53.3	66.2	25.2	544.8
1974	.3	662.9	12.2	70.0	34.4	779.9
1973	.5	870.4	22.3	25.5	8.7	927.4
TOTAL	17.4	11,903.5	747.3	7,250.6	1,806.4	21,725.2
AVERAGE	1.7	1,190.3	74.7	725.0	180.6	
% Each Species 10 Yrs.	.08%	54.8%	3.4%	33.3%	8.3%	
% OF Statewide Catch For Species, 10 Yrs.	.25%	6.3%	2.8%	1.9%	2.4%	ALL: 3.2%

In September 1982 there were 101 limited entry permits for purse seine in the Chignik district.

91

fully counted through the Chignik weir, and substantial escapement of both Black lake and Chignik Lake fish is ensured before commercial fishing is allowed. This important fishery has obviously gained: annual catches from 1973 through 1982 have seen six of the 10 years with more than 1 million of these top quality red salmon harvested: nearly 2 million were taken in both 1977 and 1981.

Pink salmon are smaller and far less valuable per pound than red salmon, and Chignik fishermen view pinks as a bonus to their bread-and-butter red salmon. During the 10 years 1973-82 an average of 725,000 pinks (33.3% of the total catch for Chignik) were annually taken in the Chignik district, both at Chignik, and in various of the many bays along the rugged 175-mile coast where there is sheltered water for laying out a seine.

It is while seining for pinks that most of the chums of the Chignik district are caught: an annual average of about 180,000 (8.3% of the total Chignik catch) were taken during the same period.

King and coho salmon are minor species at Chignik: except that, in common with many other fishing districts, in 1982 the coho catch skyrocketed. A total of 289,000 were caught. The annual average for the previous nine years was 46,000. About 1,740 king salmon are annually caught at Chignik.

Chignik salmon fishermen preparing to make a seine set. (Sharon Paul, reprinted from ALASKA® magazine)

The Alaska Peninsula — Aleutians Districts

The tapering, mountain-spined Alaska Peninsula thrusts its treeless way south and west 400 miles into the heaving North Pacific. Beyond lie 1,100 miles of the stepping stones that are the lush Aleutian Islands.

Alaska's rugged commercial salmon fishermen push their boats as far into this often-foggy, usually windy region as is commercially practical — to Unalaska Island, 200 miles beyond the tip of the peninsula. For administrative purposes the area is broken into two fisheries — the Aleutians and Alaska Peninsula. In turn, the Alaska Peninsula is broken into the North Peninsula and the South Peninsula.

Fishermen who have limited entry permits for these districts may fish the north and south sides of the Alaska Peninsula, as well as the Aleutians. Set-net fishermen work in the Shumagin Islands and Stepovak Bay (south side of the peninsula) during June and July, and at Nelson Lagoon and Port Heiden (north side of the peninsula) throughout the salmon season. The drift and seine vessels, which fish most of the areas, catch the bulk of the fish.

The Alaska Peninsula area includes the coastline from Cape Menshikof, near Ugashik, on the Bering Sea side of the peninsula, south and west along the coastline 500 miles to Unimak Pass, then for 375 miles back along the rock- and island-strewn Pacific side of the peninsula to Kupreanof Point, near the village of Perryville. Port Heiden (pop.

A little over 10 years ago, about 41 persons lived in the village of False Pass on the east end of Unimak Island. The salmon cannery was closed and there was little to stir the village's economy. The cannery was reopened in 1977, however, and now the 55 inhabitants of False Pass share in the multiple species fishing boom that has come to the eastern Aleutians. (Augie Kochuten, reprinted from ALASKA GEOGRAPHIC®)

90), and the tiny villages of Nelson Lagoon and Port Moller, are the sole human habitation centers on the north side: King Cove (472), False Pass (70), Cold Bay (226), and Sand Point (619) are found on the south side.

The Aleutian Islands area is everything west of, and including, Unimak Pass. Practically the only commercial salmon fishing in the Aleutians is for pinks, with a smattering of chums and reds, at 35-by-80-mile Unalaska Island. Unalaska (pop. 1,301) and Akutan (126) are centers of habitation on Unalaska and Akutan islands.

The first ever extensive survey of Aleutian Islands salmon streams west of Unalaska Island was completed in the summer of 1982. Moderate pink salmon runs occur on Umnak, Atka,

Adak, and Attu, with smaller runs on many of the other islands. Isolated small runs of sockeye, chum, and coho salmon were also found.

This is earthquake country, where many of the world's shudders annually occur, and where smoking mountain peaks frequently spew lava and flames. It is a land that year-round calls for good woolen clothing to be worn under rain gear. Damp cold with clouds and fog, with constant wind, bite at a man's vitals; mean January minimum temperature is a seemingly mild 28° F, while maximum July temperatures average about 56° F.

The Aleutians (Unalaska Island) fishery is an on-again off-again affair. There was essentially no fishery there from 1973-78 — fishing was more attractive elsewhere. In even years when a strong pink return is expected at Unalaska, a part of the South Peninsula seine fleet may move to Unalaska about July 20 and remain there until the end of the season, in early or mid-August.

Even-year pink runs are still dominant at Unalaska: 2.6 million pinks were caught there in 1980, and 1.4 million in 1982, with lesser catches in between. Pinks are the only important species for salmon fishermen at Unalaska.

The salmon fishery on the south side of the Alaska Peninsula is a mixed bag. Fishing usually starts in June at South Unimak or False Pass, in the Shumagin Islands, at Swedania Point, and in Stepovak Bay, where red and chum salmon bound for distant fisheries such as Bristol Bay, the Yukon River, and probably Chignik, arrive in great schools. These fish, heading to home streams after their long sojourn in the North Pacific, generally swim through known passes as they enter the Bering Sea, at times concentrating in great numbers where eager peninsula fishermen wait.

Tagging studies have shown that there is much annual variance in the proportions of red salmon arriving at South Peninsula fishing grounds and bound for Bristol Bay, North Peninsula, and local areas. In some years most are bound for Bristol Bay. Many reds taken on the eastern end of the South Peninsula district, near or at Kupreanof Peninsula, and in the Shumagin Islands, are bound for Chignik.

Such fisheries are commonly called "cape fisheries" because catches are commonly made at the ends of prominent capes, and the salmon stocks are mixed: schools may include in them stocks bound for 10 or 15 different rivers that may be hundreds of miles away. The June fishery of the South Peninsula is one of Alaska's biggest and oldest cape fisheries.

The South Peninsula June cape fishery is a mixed blessing. The shower of gold represented by millions of surging, traveling fish, bound for distant places, can make fishermen rich. But in years when, say, the Bristol Bay run is weak, valuable stock needed for spawning can wind up in a can. South Peninsula fishermen get a crack at the Bristol Bay fish before the Bristol Bay fishermen do.

But during years when 20 to 30 million or more red salmon are bound for Bristol Bay, the South Peninsula fishery relieves some of the pressure on the processors of Bristol Bay — and the South Peninsula catch provides a valuable gauge of the strength of the Bristol Bay run, as well as of the size of the fish.

The Alaska Board of Fisheries has developed a simple plan to protect those distant fisheries from overharvest by South Peninsula fishermen. In addition to a quota, escapement levels at Chignik, for example, determine fishing time for a part of the South Peninsula fishery (East Stepovak, West Stepovak, Balboa Bay and Beaver Bay sections just south and west of Kupreanof Peninsula where it is known that Chignik-bound reds are caught).

Percentages of the numbers of red salmon forecast for Bristol Bay are allowed to be harvested throughout June at South Unimak and in the Shumagin Islands where tagging has shown that Bristol Bay reds are caught in large numbers.

Catch records for the South Peninsula are often given for June, and after June, recognizing that those salmon caught in the area after June are mostly bound for local rivers and creeks.

After the June fish go by, South Peninsula fishermen have a July-early

Sand Point salmon fishermen catch many salmon bound for Bristol Bay, the Yukon and Kuskokwim rivers, and elsewhere, as well as those bound for local streams. (Lael Morgan, staff)

Harold Bendixen weighs in salmon on tender at False Pass. A cannery was established here in the 1920s, then closed in the 1970s. Today the False Pass cannery has reopened, giving a new boost to the local economy.
(Lael Morgan, staff)

August chance at pinks and chums, while North Peninsula fishermen are taking reds and chums.

The South Peninsula salmon fishery started in 1889, the North Peninsula fishery commenced about 1906, while the Aleutian fishery was under way by about 1911.

The South Peninsula salmon fishery is important, catching (during the 10 years 1973-82) about 8.3% of the salmon caught in Alaska. Annual average catch for those years was 5.5 million fish, of which 20.1% were red salmon (from local and migrating stocks), 64.1% were pinks (mostly local stocks), and 13.9% were chums (both local and migrating stocks).

Total chum salmon catches for the Peninsula-Aleutians area generally ran from 1 to 2 million fish annually from 1924-1960, then decreased. During the 10-year period 1973-82, the annual average catch was around 1 million, nearly 80% of which were caught in the South Peninsula. These chums were headed for both local spawning systems and distant areas. A small tagging study in June 1961 indicated that some of the migrating fish caught were bound for North Peninsula streams, and some for Western Alaska streams (Bristol Bay, the Kuskokwim and Yukon; some may be bound for Asia, most likely the USSR).

In 1982 the Alaska Board of Fisheries adopted a policy of non-expansion of the chum salmon catch rate for the South Peninsula, and directed the Department of Fish and Game to conduct studies on stock separation. There is some indication that in 1982 South Peninsula fishermen intercepted considerable numbers of chum salmon bound for the Yukon and Kuskokwim rivers.

Little is known of chum salmon in the Aleutians, although a small run exists at Massacre Bay on Attu Island.

A peculiarity of the red salmon fishery of the South Peninsula is the occasional appearance, between July 5 and 20, of many immature red salmon in the Shumagin Islands headland fishery. Seiners hauling their nets have suddenly found these young red salmon gilled in the 3.5-inch-mesh seine — as many as 1,000 in a 20-minute set. This has occurred several times in recent years. When this happens the salmon fishery is closed in order to protect these immature fish, which have no market value. Most fishery scientists think they are most likely wandering Bristol Bay red salmon.

The North Peninsula fishery, much smaller than that of the south side, depends mostly upon local stocks of sockeye and chums. Average annual catch 1973-82 was 1.3 million, of which 69.3% were red salmon, bound mostly for Nelson Lagoon, Bear Lake, Ilnik, and other local systems.

Salmon catch for
Aleutian District

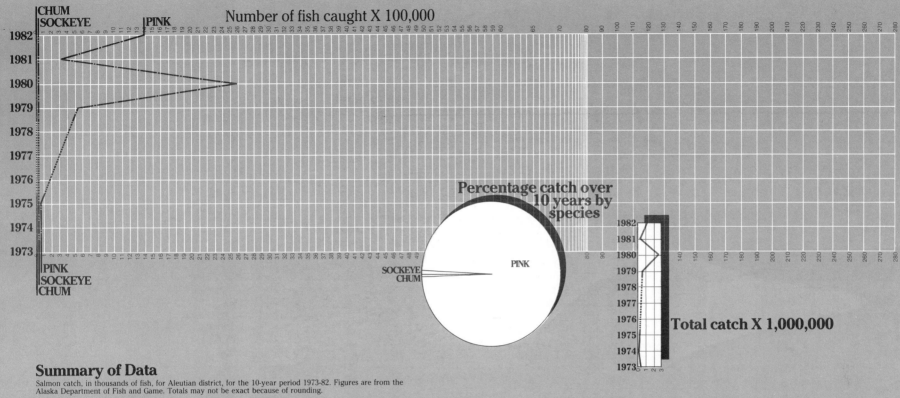

Number of fish caught X 100,000

Percentage catch over 10 years by species

Total catch X 1,000,000

Summary of Data

Salmon catch, in thousands of fish, for Aleutian district, for the 10-year period 1973-82. Figures are from the Alaska Department of Fish and Game. Totals may not be exact because of rounding.

YEAR	CHINOOK	SOCKEYE	COHO	PINK	CHUM	TOTAL
1982	—	3.0	—	1,390.0	4.0	1,397.0
1981	—	5.4	.2	302.8	6.6	315.0
1980	—	9.2	—	2,597.5	4.9	2,611.6
1979	—	12.2	—	539.4	.2	551.8
1978	No fishery					
1977	No fishery					
1976	No fishery					
1975	—	19.4	—	.7	1.9	21.9
1974	—	—	—	—	.1	.2
1973	—	—	—	2.8	—	2.8
TOTAL	—	49.2	.2	4,833.2	17.7	4,900.3
AVERAGE	—	4.9	—	483.3	1.7	490.0
% Each Species 10 Yrs.		1.0%		98.6%		
% Of Statewide Catch For Species, 10 Yrs.				1.2%		ALL: .7%

Salmon catch for
North Alaska Peninsula Fishing District

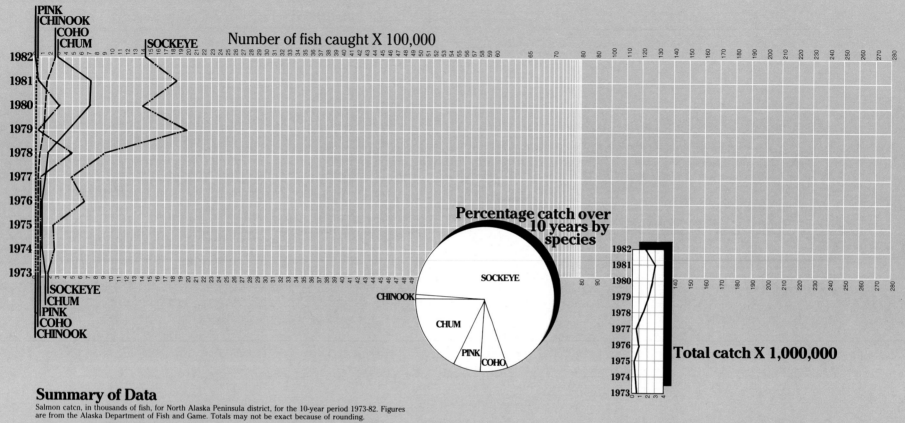

Number of fish caught X 100,000

Percentage catch over 10 years by species

Total catch X 1,000,000

Summary of Data

Salmon catch, in thousands of fish, for North Alaska Peninsula district, for the 10-year period 1973-82. Figures are from the Alaska Department of Fish and Game. Totals may not be exact because of rounding.

YEAR	CHINOOK	SOCKEYE	COHO	PINK	CHUM	TOTAL
1982	30.0	1,418.0	240.0	9.0	286.0	1,983.0
1981	15.8	1,822.9	155.4	11.2	706.4	2,711.8
1980	16.8	1,397.4	127.9	301.7	700.3	2,544.1
1979	17.1	1,979.5	112.8	5.2	65.7	2,180.1
1978	14.3	896.9	63.3	485.2	163.9	1,623.6
1977	5.5	473.1	34.1	.9	129.5	643.1
1976	5.0	642.5	26.1	.7	74.0	748.1
1975	2.1	233.3	28.4	.3	8.8	272.8
1974	5.3	255.5	24.2	10.6	35.6	331.2
1973	4.6	172.1	26.9	.4	155.8	359.8
TOTAL	116.5	9,291.2	839.1	825.2	2,326.0	13,397.6
AVERAGE	11.6	929.1	83.9	82.5	232.6	1,339.7
% Each Species 10 Yrs.	.86%	69.3%	6.2%	6.1%	17.3%	
% Of Statewide Catch For Species, 10 Yrs.	1.7%	4.9%	3.2%	.2%	3.1%	ALL: 2.0%

Salmon catch for
South Peninsula District

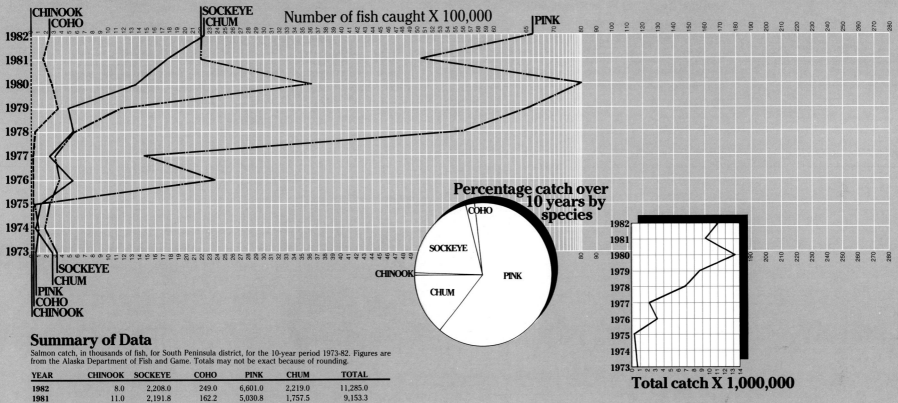

Number of fish caught X 100,000

CHINOOK
COHO
SOCKEYE
CHUM
PINK

SOCKEYE
CHUM
PINK
COHO
CHINOOK

Percentage catch over 10 years by species

COHO
SOCKEYE
CHINOOK
CHUM
PINK

Total catch X 1,000,000

Summary of Data

Salmon catch, in thousands of fish, for South Peninsula district, for the 10-year period 1973-82. Figures are from the Alaska Department of Fish and Game. Totals may not be exact because of rounding.

YEAR	CHINOOK	SOCKEYE	COHO	PINK	CHUM	TOTAL
1982	8.0	2,208.0	249.0	6,601.0	2,219.0	11,285.0
1981	11.0	2,191.8	162.2	5,030.8	1,757.5	9,153.3
1980	4.8	3,613.0	274.2	7,861.5	1,353.1	13,106.6
1979	2.1	1,149.9	356.9	6,564.9	482.9	8,556.8
1978	.8	579.4	60.8	5,590.1	546.2	6,777.3
1977	.6	311.7	2.1	1,448.6	243.2	2,006.2
1976	2.2	375.0	.2	2,366.8	532.5	3,276.8
1975	.1	243.5	.1	60.6	130.8	435.1
1974	.6	197.4	9.4	100.6	71.8	379.8
1973	.4	330.1	6.6	58.1	292.9	688.1
TOTAL	30.6	11,199.8	1,121.5	35,683.0	7,629.9	55,665.0
AVERAGE	3.0	1,119.9	1,12.1	3,568.3	762.9	5,566.5
% Each Species 10 Yrs.	.05%	20.1%	2.0%	64.1%	13.7%	
% Of Statewide Catch For Species, 10 Yrs.	.45%	5.9%	4.2%	9.4%	10.4%	ALL: 8.3%

Limited entry permits held for South Peninsula, North Peninsula, and Aleutian districts (all are open for the same permit) in September 1982 included 127 purse seine, 164 drift gill net, and 116 set gill net.

The Bristol Bay District

Bristol Bay is the most valuable salmon fishery in Alaska. Fishermen catch an incredible number of salmon there each year — more red salmon than are taken in any other fishery in the world: for 20 years the annual catch has averaged more than 10 million silver-sided ruby-fleshed reds. During the 10 years 1973-82 the annual average sockeye catch was 11.4 million. For the years 1973-82, Bristol's Bay's fishermen caught 61.1% of all the sockeye salmon caught in Alaska. All species considered, during this 10-year period Bristol Bay produced one-fifth (20.6%) of all the salmon caught in Alaska.

But when you talk about the bay (to Alaska's commercial salmon fishermen "the bay" means only one place — Bristol Bay) you're talking about red or sockeye salmon, which is the choice of fishermen because, except for the far less abundant king salmon, it brings more per pound than any of the other species of salmon. It's the choice of packers because of the rich, firm, red flesh, for which consumers are willing to pay premium prices, whether fresh, frozen, or canned.

What do 11 million salmon look like? If 11 million — the recent average annual catch of Bristol Bay's 24-inch-long red salmon — were placed head to tail, they would stretch from New York City to San Francisco — and then on to Albuquerque, New Mexico (4,154 miles), and a few miles more.

This incredible catch is almost explosive, for Bristol Bay's sockeye season is one of the shortest in Alaska: the fish generally start to arrive during the last week of June, the peak of the run occurs about July 4, and it is essentially over by mid-July — just over two weeks. Up to 1.6 million fish have been caught in the bay **during one 12-hour fishing period in a single district**.

This stupendous catch is made entirely by gill nets: there are 958 set gill net permits, and 1,824 drift gill nets for the bay.

Freshly gill-netted Bristol Bay sockeye. When salmon near spawning grounds they may start to develop colors and form they acquire at spawning time. Some of these sockeye have started to turn reddish. (Freda Shen)

Bristol Bay, at the base of and on the north side of the great thrusting Alaska Peninsula, is shallow, often stormy — the easternmost lobe of the Bering Sea. Most inshore areas are muddy. Huge tides (20 feet and more) push millions of tons of muddy water in and out of rivers' mouths, alternately flooding the muddy banks or exposing great bars and mud flats. Jumbled masses of ice plug these river mouths in winter.

This less-than-attractive bay, which can be reached from other parts of Alaska only by air or sea, is rimmed for 200 miles on the north and east by lovely, clear-water gravel-bottomed lakes and winding streams which hold a great wealth of all five species of Pacific salmon.

Tiny villages and small towns dot the map of Bristol Bay wherever one of these lovely lake-draining rivers reaches the sea. Most southerly is the **Ugashik district:** Upper Ugashik Lake (17 miles long) is connected by a short stream to Lower Ugashik Lake (11 miles across). They're in tundra country — west of the spruce line, and the land around them is hummocky, grass-and-alder grown. The two lakes are drained by 43-mile-long Ugashik River, which pours its crystal waters into Ugashik Bay. Pilot Point (pop. 72) is built on the shores of this small Bering Sea coast bay.

The Ugashik catch has been erratic. Best years in the past were 1941-55 with an average annual catch of 715,000 reds. During the early 1980s this run has strengthened. The 1981 catch of 2 million was an all-time record. Average catch 1980-82 was 1.3 million. In 1982 fishermen netted 1.2 million, while another 1.2 million splashed and swam up the Ugashik River to the lakes to spawn.

The Egegik fishing fleet on the mud flats at low tide. (Ken Hunter, reprinted from ALASKA® magazine)

South Naknek, center of the cannery and village complex that makes up the heart of the great Naknek-Kvichak salmon fishery. (Tom Walker)

Next up the coast is the **Egegik district.** Becharof Lake, gravel-bottomed, clear — and Alaska's second largest, is 15 miles wide, 37 miles long. The winds that commonly whoop off of the Bering Sea can turn this huge lake into a veritable maelstrom. But its clear water and gravelly bottom are ideal for spawning and rearing of red salmon. Twenty-eight-mile-long Egegik River drains this huge lake and pours the clean waters into Egegik Bay where lies the tiny wind-battered coastal village of Egegik (pop. 75). The all-time high catch for this district was in 1981 when fishermen netted 4.5 million sockeye. In 1982, 1 million sockeye salmon swam past this village, up the river, and into Becharof Lake: fishermen caught another 2.4 million in Egegik Bay.

About 40 miles to the north the **Naknek** River pours its gin-clear water into Bristol Bay, after draining 10-by-20-mile Naknek Lake, and several others in the drainage. This is spruce country, and stands of small spruce are found generally around the Naknek system. At the mouth lies Naknek (pop. 317), and, on the south side of the river mouth, South Naknek (pop. 147). Upstream, connected to Naknek by a 15-mile road, is King Salmon (pop. 317).

Linked with the Naknek River and its great run of sockeye is the biggest producer of all — the Kvichak. Commonly they are lumped. The **Naknek-**

Kvichak district they're called. The winding, sometimes shallow, clearwater 50-mile-long Kvichak River drains Lake Iliamna, Alaska's largest. Appropriately, this system produces more red salmon than any other, anywhere. Over the years about 63% of all red salmon caught in Bristol Bay have been taken in the Naknek-Kvichak fishing district, with about 85% of the district run being fish bound for the Kvichak.

Sockeye runs to Iliamna Lake are extremely cyclic: peak years are separated by years of lower production. In recent years the peaks have occured on a bi-decade basis — i.e., 1965, 1970, 1975, 1980. A "variable-cycle year escapement strategy" for the Kvichak River system developed by state fishery scientists is designed for greater production spread over more years, and the system appears to be entering a new and different phase of salmon production.

Small villages are scattered around huge Iliamna Lake, including Pedro Bay (pop. 42), Iliamna (94), Igiugig, Kokhanok, Pile Bay, Newhalen. Levelock (80) is a tiny old village on the banks of the Kvichak River near where it enters salt water.

About 45 miles due west of Naknek is Etolin Point, and 30 miles southwest of that is the tip of the Nushagak Peninsula: between the two is the great, deep, narrow **Nushagak Bay and**

Naknek Harbor, at night, during the frenzied peak of the Bristol Bay sockeye run. (Freda Shen)

district, second largest producer of red salmon in the bay. Dillingham (pop. 1,535), largest community in the region, lies on the shores of this bay. Just north of Dillingham is found the secret to the great salmon production of this bay: two systems of five lakes each making up the loveliest, clear-water, deep lakes found anywhere.

The Wood River Lakes lie closest to Dillingham — and the nearest of those, Lake Aleknagik, is connected to Dillingham by road. All five are connected by short rivers, and all are drained by Wood River, which pours into Nushagak Bay near Dillingham.

Just above the Wood River Lakes are the Tikchik Lakes, similarly lovely, clear-water and deep, nestled amidst spruce-flanked mountains. All are connected by short streams. They drain into the Nushagak River upstream from Wood River. So beautiful is this region that Alaska has made it the Wood River-Tikchik Lakes State Park.

Villages dot the shores of Nushagak Bay and River — including Clarks Point (pop. 79), Ekwok (79), New Stuyahok (325), and Koliganek (116). Aleknagik (154) is located at the lake by that name.

During early years of the salmon fishery the Nushagak district annually produced 5 million (1899-1918), 2.8 million (1919-48), and thereafter until

The Standard Oil dock and the Peter Pan salmon cannery at Dillingham, Bristol Bay, is a busy spot during the summer sockeye run.
(John and Margaret Ibbotson)

the late 1970s (1949-77), 882,000 red salmon per year. During the last few years of the 1970s and in the early 1980s the Nushagak has again come to life: the 1981 catch of 7.7 million was the largest ever. In 1982 fishermen netted 6 million red salmon in the Nushagak, the seventh largest catch ever for that system. Escapement was 2.9 million in 1981, 2 million in 1982. Fishery scientists say that smolt production has been high in the Nushagak in recent years, and the near future, at least, looks bright.

About 65 miles to the west of Nushagak Bay lies **Togiak Bay and district**, into which drains the Togiak River — which in turn drains Togiak Lake and Upper Togiak Lake, another fine red salmon-producing system. Smaller than the other systems, red salmon catches here range around 200,000, although in 1982, the catch was 584,000, the third largest ever, with escapement of 341,000. The village of Togiak (pop. 472) lies on the shores of the bay and on the banks of the Togiak River.

These then, are the five fishing districts of the bay. Their existence depends upon rich lakes that seem to have been designed for producing vast, endless numbers of salmon. Annually, millions upon millions of silver-sided strong-swimming red-fleshed salmon pour into the rivers that drain these lakes. Figures are a poor way to describe their numbers.

Each June, Bristol Bay-bound sockeye salmon pour through Unimak Pass and other passes in the Aleutian Islands. Noses pointing like the needle of a compass, they head for Bristol Bay in huge schools and long strings. By the time they are opposite Port Moller, on the north side of the Alaska Peninsula, and within 300 miles of the bay, they are swimming within an 80-mile-wide band next to the shoreline. As they near the bay, they concentrate offshore in the southern half of the entrance to the bay, and then in the southern half of the bay itself.

Bringing in a cripple. This drift gill-netter had engine trouble, and was towed across Nushagak Bay to Dillingham by the tender Baranof. *Wake of the* Baranof *threatened to swamp it a time or two. Salmon tenders are mother hens to the drift gill-net fleet, providing fuel, food, water, parts, and emergency rescue when necessary. (Jim Rearden, staff)*

The Peter Pan cannery tender **Baranof** *with a deckload of sockeye salmon, Nushagak district, Bristol Bay. Tenders like the* **Baranof** *are short-term handlers: fish are loaded and unloaded within hours, and the* **Baranof** *makes short runs. Weather is normally cool, and the fish remain fresh and firm. Tenders used for longer hauls, or that hold fish for longer periods, use chilled brine tanks to hold the salmon.* (Jim Rearden, staff)

Segregation according to river of origin apparently starts in the offshore waters as much as 120 miles from the mouths of the home-river systems. The steadily swimming fish remain offshore until they are within 19 to 48 miles of their home-river systems. Once they have successfully navigated the muddy, windy, and restless bays — the Ugashik, Egegik, Naknek-Kvichak, Nushagak, Togiak — the fish swim steadily upstream in tight bands along the riverbanks: you can fly over them and peer down into the clear water and watch the undulating lines of fish, mile upon mile of them, as they steadfastly move toward the place of their birth. It's one of nature's grandest spectacles.

The total inshore sockeye salmon catches in Bristol Bay for the period 1893-1971 annually averaged 11.39 million; however, from 1956 through 1971 Japanese high-seas fishermen caught an additional 39.5 million Bristol Bay-bound sockeye. Including the Japanese catch, the average per year harvest for Bristol Bay-bound sockeye for 1893-1971 was 11.89 million — more than 4,154 miles of nose-to-tail salmon.

Back-to-back severe winters in 1970-71 and 1971-72 froze eggs in the gravel, created icy sea temperatures where plankton couldn't grow, and wiped out most of those year classes. Catches during the 1972-77 rebuilding period dropped to the all time low of only 3.3 million per year. Fishermen sat on the beach while valuable brood stock swam through the net-free bay to safely reach spawning grounds.

The sacrifice paid off starting in 1978 when the depressed fishery started to rebound; the red salmon catch that year was 9.9 million. Average annual catch of sockeye in the bay, 1979-82, was 23.2 million.

King salmon make up less than 1% of the total bay salmon catch (1973-82), yet the Bristol Bay catch for those years was 19.9% of the statewide total. The annual average king catch for the bay for that period was about 134,000. Recent years have seen an upsurge in king salmon runs. In 1982, the catch of the big spotted kings in the bay totaled 253,000.

The Nushagak district produces most of the kings in the bay. Set-net and drift fishermen usually have their nets out by early June, waiting to see their corks dip as the 20-pounders hit. In 1982, for example, 200,000 kings were caught by fishermen in the Nushagak. Another 40,000 were caught by Togiak district fishermen, and the balance of the 253,000 in other districts of the bay.

Chum salmon catches for the bay 1973-82 averaged about 1 million annually, most of which were caught in the Nushagak district. Chums arrive during the sockeye salmon run, and are a "me too" fish for the bay: it's not practical to shut down a multi-million dollar sockeye salmon fishery to ensure that a relatively few chum salmon reach spawning areas. Fishermen catch them in their red salmon nets.

In 1982 the chum catch for the bay was about 936,000, of which 456,000 were caught in the Nushagak district.

Pink salmon are important only in

the Nushagak district, and then only in even-numbered years. Prior to 1918 there was an odd-year pink run, but for unknown reasons, that run disappeared.

Pink salmon supplements the red salmon catch mostly for local residents — fishermen who remain after the peak of the great red salmon run has passed, for that is when the pinks arrive in numbers.

The Nushagak pink run fluctuates considerably. In 1978, for example, 13.7 million pinks returned, and 5.1 million of them entered the commercial catch in the Nushagak district. The 1980 catch was down to 2.5 million, while the 1982 catch dropped to 1.4 million. Odd-year catches, 1977, 1979,

The old and new. Outside boat was at one time a Bristol Bay sailboat, fished prior to 1953 when sailboats only were allowed to fish for salmon in Bristol Bay. Mast and sailing gear have been removed, an engine and small cabin installed, and it is here used as a power boat.

Most of the "conversions" as they were called (converted from sail to power) are now gone, replaced by such boats as the one shown next to the conversion. Most of the new boats are fiberglass or aluminum, many are diesel powered. Most are beamy and capable of good speed. Maximum length of a drift gill-net vessel in Bristol Bay is 32 feet: it's illegal to fish one that is longer. (Jim Rearden, staff)

Spawning red or sockeye salmon in Naknek Lake, Bristol Bay drainage.
(Tom Walker)

and 1981 were respectively, 3,000, 2,000, and 300.

The annual even-year average pink catch for the bay, 1974-82 (five years) was 2.2 million; 1982, was 1.2 million fish.

Coho salmon are also caught mostly by local fishermen who live around the bay, for they arrive last, long after the great sockeye salmon runs have ended. Often the coho salmon gives the hard-luck fishermen a payday — the fisher-men whose boat engine blew up during the peak of the red season, or who for one reason or another didn't get in on the gravy.

For the 10-year period 1973-82 the annual average catch of coho salmon in Bristol Bay was 205,000. Since large numbers of fishermen have abandoned the bay by the time the cohos arrive, the species has special significance to the relatively few fishermen who share in this catch, even though it repre-sented only about 1.4% of the total number of fish taken in Bristol Bay (1973-82).

The Nushagak district produces the largest number of cohos, with the Togiak district following.

In common with most other regions of Alaska, the 1982 coho catch for Bristol Bay was spectacular, totaling 654,000 — more than twice the biggest catch of any of the 10 previous years.

Salmon catch for
Bristol Bay

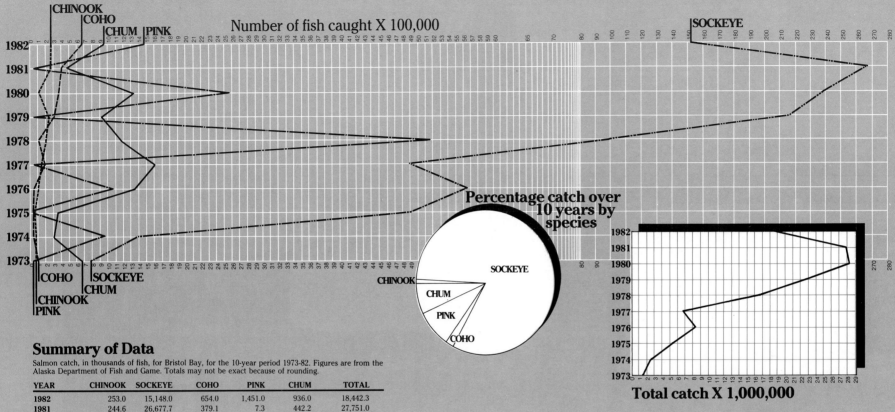

Number of fish caught X 100,000

CHINOOK
COHO
CHUM PINK
SOCKEYE

COHO   SOCKEYE
              CHUM
CHINOOK
PINK

Percentage catch over
10 years by
species

SOCKEYE
CHINOOK
CHUM
PINK
COHO

Total catch X 1,000,000

Summary of Data

Salmon catch, in thousands of fish, for Bristol Bay, for the 10-year period 1973-82. Figures are from the Alaska Department of Fish and Game. Totals may not be exact because of rounding.

YEAR	CHINOOK	SOCKEYE	COHO	PINK	CHUM	TOTAL
1982	253.0	15,148.0	654.0	1,451.0	936.0	18,442.3
1981	244.6	26,677.7	379.1	7.3	442.2	27,751.0
1980	95.5	23,762.7	348.5	2,563.5	1,300.0	28,070.3
1979	212.9	21,428.6	294.4	3.8	906.8	22,846.5
1978	191.5	9,928.1	94.3	5,152.7	1,158.1	16,524.7
1977	130.5	4,877.9	107.2	4.5	1,598.2	6,718.5
1976	96.0	5,619.3	26.6	1,036.5	1,329.1	8,107.5
1975	30.0	4,898.8	46.3	.4	325.4	5,300.9
1974	45.7	1,362.5	43.7	940.0	286.4	2,678.2
1973	44.0	761.3	57.0	.4	648.5	1,547.3
TOTAL	1,344.0	114,464.9	2,051.1	11,160.1	8,966.7	137,987.2
AVERAGE	134.4	11,446.4	205.1	*1,116.0	896.6	13,798.7
% Each Species 10 Yrs.	.9%	82.9%	1.4%	8.0%	6.4%	
% Of Statewide Catch For Species, 10 Yrs.	19.9%	61.1%	7.8%	2.9%	12.3%	ALL: 20.6%

*Average for all years. Even year average (1974, 1976, 1978, 1980, 1982) is 2,228.7.
Limited entry permits held for Bristol Bay in September 1982 included 958 set gill net, and 1,824 drift gill net.

The Kuskokwim District

The 850-mile-long, mostly silty, winding, sluggish Kuskokwim River drains 4,500 square miles of spruce, birch, and tundra land of Western Alaska. Second only to the Yukon River in Alaska in size and length, this great river originates in central Interior Alaska near the village of Medfra and flows southwest through the wide, flat, pond-and-lake-filled Kuskokwim valley to empty into the Bering Sea. East of the river for most of its length lie the low but rugged Kuskokwim Mountains.

About 10,000 people live in the Kuskokwim Basin, 3,500 of them in Bethel, on the lower river. The rest live in dozens of small riverbank villages scattered throughout the area. The upriver village of McGrath (355) and the mid-river village of Aniak (341) are other major centers of commerce.

No roads extend into the Kuskokwim valley: local roads exist around Bethel and other villages. Transportation is mostly by air. River transportation is possible for about five months: the rivers are frozen over during the rest of the year. Climate is coastal to continental, with deep cold winters and warm summers inland, and blustery, moderately wet, often cloudy, foggy days along the cool Bering Sea coast.

Kuskokwim River valley residents have always depended upon the food provided by annual runs of all five

Hauling in king salmon on the Kuskokwim River near Bethel.
(James Barker)

species of salmon. Prior to statehood, no export of salmon was permitted from the Kuskokwim River, following a furor over operations of canners in the early 1920s. The no-export policy limited salmon production of the area to a couple of small hand canneries and some salting for Alaska sale only. The restriction was lifted after statehood and with generally increasing scarcity of choice king and other salmon species, commercial operators increased their attention to Yukon River (where the restriction was also in effect) and Kuskokwim fish, and changed the Eskimo and Indian subsistence economy into a cash business. The side-by-side Kuskokwim and Yukon rivers are the only rivers in the state where commercial salmon fishing is allowed.

The Kuskokwim commercial salmon fishery is the oldest in the Arctic-Yukon-Kuskokwim region, with catches reported as early as 1913. But commercial fishing didn't really mature for about half a century. For years small commercial mild-cure (salting) operations were conducted in Kuskokwim Bay and at the mouth of the Kuskokwim River, while the Kuskokwim River itself remained virtually untouched.

In the 1930s, when Alaskans depended upon dog teams for hauling freight during the winter, a quasi-commercial fishery existed in the McGrath area for the sale of dried subsistence-caught salmon for dog food. When airplanes largely replaced dog teams for winter freight hauling, this fishery died out, and there was little commercial use of the Kuskokwim's salmon until shortly after statehood.

Now fishermen are amidst the transition from subsistence to commercial fishing, and the lives of many Kuskokwim valley residents are a blend of the two. Subsistence is still the priority use for the Kuskokwim salmon, but the commercial use has greatly increased. In this region, subsistence and commercial fishermen are the same individuals, and commonly the same nets are used for both kinds of fishing.

In 1982 there were 837 salmon limited entry permits for salmon gill nets for the Kuskokwim Bay and River. Permits may be used to fish set gill nets or drift gill nets, both of which are permitted in all districts. The trend is toward more drift gill nets, for they are more efficient.

Commercial salmon fishing is allowed along 250 miles of the lower Kuskokwim River (to the Kolmakoff River, about 21 miles upstream from Aniak), and in the Quinhagak and Goodnews Bay subdistricts located along the coast in Kuskokwim Bay.

Although the commercial salmon harvest for the Kuskokwim district is small in relation to the rest of the state (less than 1%) it is extremely important to the mostly-Yup'ik Eskimo fishermen, who have few sources of income other than commercial fishing.

For the 10-year period 1973-82, the average annual total catch of salmon in the Kuskokwim district was 677,200. Chums were the most numerous (46%), followed by cohos (38%), kings (8%), sockeye (5%) and pinks (3%). Most of the catch is fresh-frozen for the market.

Chum salmon spawn in at least 16 tributaries of the Kuskokwim River, several of which are more than 75 miles long. In the Kuskokwim Bay area they use most of the Kanektok River, the lower part of the Arolik River, Jacksmith Creek, Cripple Creek and Indian River. Chums do not appear to be cyclic in the Kuskokwim, although the runs do fluctuate considerably in size. For the years 1973-82 an annual average of 308,100 chums were netted for commercial sale from the Kuskokwim.

Coho salmon run in August when stream levels are high and muddy, making it difficult to determine spawning areas and to count spawners. This is also the time when the people are berry picking, getting ready for moose hunting, and other pre-winter food gathering occupations. Weather is wet, and not suitable for drying salmon, and in the past coho salmon were not of great importance in the subsistence fishery. With freezer facilities available in most villages now, cohos are steadily growing in importance for this fishery.

Coho salmon use many of the same rivers for spawning as do the king salmon. Included in known spawning systems for coho are the upper parts of the Kwethluk, Aniak, Salmon, Kipchuk, Oskawalik and Kogrukluk rivers.

In common with most fishing districts in Alaska, 1982 was a boom year for cohos in the Kuskokwim: happy Eskimo fishermen landed

A commercial fisherman watches carefully as his catch is weighed by a fish buyer on the Kuskokwim.
(James Barker, reprinted from ALASKA GEOGRAPHIC®)

569,000 of them in that year — the largest coho catch on record. The recent 10-year average annual catch was 254,700.

Pink salmon runs into the Kuskokwim are strongest in even years. Limited numbers of pinks spawn in various mostly downriver tributaries of the Kuskokwim — the upper Kwethluk and Holitna rivers, and lower parts of the Kogrukluk. Large runs of pinks occur in the Kanektok and Arolik rivers and in upper Goodnews River. Even-year catches for the past 10 years have ranged from 18,000 to 62,000. Odd-year catches of pinks are insignificant.

King salmon spawn in various tributaries of the Kuskokwim, including the Kwethluk, Kisaralik, Tuluksak, Aniak, Salmon, Kipchuk, Holitna, Kogrukluk, Hoholitna and Chukowan rivers.

Most of the kings passing through the Quinhagak fishery spawn in the Kanektok River, where spawning densities are among the highest of the area. King salmon have also been seen spawning in the Arolik River.

During the 10 years 1973-82 the annual average king salmon catch for the Kuskokwim was 53,800, which is about 7% of the total state catch of kings for that period.

Salmon catch for
Kuskokwim Fishing District

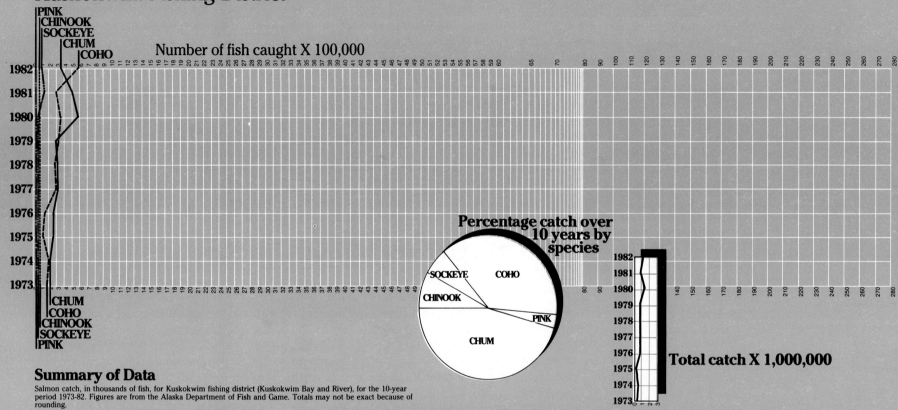

Number of fish caught X 100,000

PINK
CHINOOK
SOCKEYE
CHUM
COHO

1982
1981
1980
1979
1978
1977
1976
1975
1974
1973

CHUM
COHO
CHINOOK
SOCKEYE
PINK

Percentage catch over 10 years by species

SOCKEYE
COHO
CHINOOK
PINK
CHUM

Total catch X 1,000,000

1982
1981
1980
1979
1978
1977
1976
1975
1974
1973

Summary of Data

Salmon catch, in thousands of fish, for Kuskokwim fishing district (Kuskokwim Bay and River), for the 10-year period 1973-82. Figures are from the Alaska Department of Fish and Game. Totals may not be exact because of rounding.

YEAR	CHINOOK	SOCKEYE	COHO	PINK	CHUM	TOTAL
1982	79.7	97.6	569.0	18.1	325.1	1,089.5
1981	79.3	106.5	278.3	.5	482.5	947.1
1980	48.8	42.6	327.6	29.9	558.9	1,007.7
1979	53.7	39.5	304.2	.6	295.8	693.8
1978	64.5	14.0	246.8	62.3	282.2	669.7
1977	58.8	18.6	263.9	.5	297.8	639.6
1976	50.1	14.5	112.3	40.4	232.6	449.9
1975	27.8	17.5	109.8	.9	223.5	379.6
1974	29.8	27.9	176.5	60.3	193.4	387.9
1973	51.4	5.2	152.4	.6	184.2	393.8
TOTAL	543.9	383.9	2,540.8	214.1	3,076.1	6,758.6
AVERAGE	54.3	38.3	254.0	21.4	307.6	675.8
% Each Species 10 Yrs.	8.0%	5.6%	37.5%	3.1%	45.5%	
% Of Statewide Catch For Species, 10 Yrs.	8.0%	.2%	9.7%	—	4.2%	ALL: 1%

Limited entry permits held for the Kuskokwim may be used either to fish set gill nets, or drift gill nets. in September 1982 there were 837 limited entry permits for this district.

The Yukon District

The broad, winding, muddy Yukon is one of the great rivers of the world. Fifth largest in North America, it is Alaska's largest river and it drains 330,000 square miles of Alaska (nearly 35% of the state) and Yukon Territory, Canada. It flows from headwaters in British Columbia, within 30 miles of the Gulf of Alaska, and rolls in a great 2,300-mile arc through Yukon Territory and Alaska, to its several broad mouths on the Bering Sea.

Commercial fishing is permitted on the main Yukon and one of its main tributaries, the Tanana: tributary streams of the Yukon and Tanana are closed to commercial fishing.

Most of the Yukon River proper is muddy, but hundreds of its tributary streams are clear and gravel-bottomed, ideal for the five species of Pacific salmon that spawn in them.

Yukon River salmon are important mainly for commercial and subsistence use to 10,000 to 15,000 Eskimos, Indians and whites who live in more than 45 small villages along the coast and along the Yukon and its tributaries. The Yukon River commercial salmon catch was about 1.6% of the total statewide catch for the 10 years 1973-82.

Some Yukon River chinook salmon must swim 2,300 miles to headwaters spawning grounds, and the energy to make this long journey comes from the great quantity of oil stored in their red flesh, for they cease eating when they enter the river on their last journey.

In one of the most complex salmon fisheries in Alaska, and by far the longest geographically, headwaters-bound Yukon River salmon must run a gauntlet of 1,200 miles of gill nets and fish wheels fished by about 950 mostly Eskimo and Indian fishermen (there are 166 limited entry permits for fish wheels for the upper Yukon, and 76 gill net entry permits: additionally, there are 710 gill net entry permits for the lower Yukon).

The Yukon River snakes its way across the lake-dotted flat tundra of the Yukon River delta, where the blustery climate of the Bering Sea dominates, thence inland through the spruce-grown uplands of Interior Alaska where there is little precipitation, with warm summers and deep cold winters.

Roughly 65,000 people live in the Yukon valley in Alaska, including 54,000 at Fairbanks. The remainder are scattered throughout the area in small riverbank villages. The Alaska Highway system is the main transportation system in the Fairbanks area, while airplanes provide most transportation for the rest of the basin. River traffic is active for only five months: the rivers are frozen over the remainder of the year.

In recent years the number of commercial fishermen of the upper river has increased. Further, subsistence is the priority use: by law, all subsistence needs must be satisfied before commercial fishing can be allowed. Lower river fishermen get first chance at the fish, but the season length for the lower river must be set so that enough fish swim by for the upriver subsistence and commercial users, as well as for spawning needs.

State fishery scientists estimate that a king salmon bound for the upper Yukon is exposed to about 25 fishing days as it swims past nets and fish wheels bound for its home stream.

Managing so that upstream fishermen get their fair share is a touchy business on the Yukon, and to accomplish it an elaborate system of fishing times, gill-net mesh sizes, and quotas, are in place for the six districts and 10 subdistricts along the river. The quotas ("guideline harvest ranges" in the jargon of fishery managers) for each district and subdistrict are annually set by the Alaska Board of Fisheries in an attempt to equitably share salmon among upriver, downriver, and middle river fishermen, with enough left for spawning. Quotas are flexible, and allow for small, large, or average runs.

Downstream fishermen are mostly Eskimo, while upstreamers are mainly Indian (only about 10% to 15% of the fishermen are not one or the other). Rivalry, suspicion, fear, and distrust marked relationships between Indians and Eskimos for centuries. But the rebirth of pride in Native heritage in Alaska with the Alaska Native Claims Settlement Act of 1971 has drawn the two races together, and in recent years

Untangling a king salmon from a set net off Emmonak keeps a father-son fishing duo busy. (Gary Foster, reprinted from ALASKA GEOGRAPHIC®)

reasonably amicable relationships have developed, at least partly because of the common interest in seeing a sustained and healthy salmon fishery on their common fishing grounds — the Yukon River.

There are also international implications, for about 40% of the Yukon River flows through Canada. Yukon Territory fishermen have complained that they do not get their fair share of Yukon River fish.

King salmon have long been the most sought after and the most valuable individual salmon of the Yukon River. Rich because of their high oil content, and with uniform red flesh — there are no white-fleshed kings here as are found in Southeastern — these 20-pound-average salmon are regarded as the finest in Alaska. A commercial fishery has existed for them in Alaska since 1918, and there are good records of Yukon king catches back to the earliest years of this century.

King salmon arrive at the mouth of the Yukon River in late May and they continue to arrive until July. It requires about 60 days for a king salmon to swim from the mouth of the river to the upper tributaries.

King salmon spawn throughout the Yukon valley in at least 56 streams in Alaska, and 52 streams in Yukon Territory. The Yukon River king salmon run is, for an individual river, second in

Ada Lawrence uses an* ulu *to cut strips of king salmon for drying racks at Mountain Village. *(Myron Wright, reprinted from ALASKA® magazine)*

size only to that of the Columbia River. During the 10-year period 1973-82, the annual average Yukon River commercial king catch was 107,000, or about 10% of the total statewide catch of the species. Biggest catch was in 1981 when 158,000 were netted.

Chum salmon are the most abundant salmon species in the Yukon River. The huge runs of chum were once mainly important as food for sled dogs when most winter transportation in Interior Alaska was by dog team. "Dogs," these salmon were called, were caught summer and fall, then split and dried, and tied into easy-to-handle bales. For years, into the 1950s, these bales or bundles of dried dog salmon were a common article of commerce with merchants and traders throughout Interior Alaska.

Dog teams were replaced largely by airplanes and snow machines. The demand for chums as dog food largely disappeared for a time, but today it is increasing because of the growing use of recreational sled dog racing teams.

Commercial use dominates today as these chums are eagerly sought by buyers who market them mostly as frozen or fresh fish. Modern air transport, freezer ships and plants, and development of Japanese markets, have made this possible. Today the Yukon chum rivals the king salmon in economic value.

The Yukon River chum salmon run is among the largest chum runs in the world. For the 10-year period, 1973-82, the annual average chum catch was 971,600 fish, or 88.3% of the average

1.1 million total salmon catch for the Yukon. Highest catch was made in 1980, when 1,473,100 chums were caught and marketed.

There are two types of Yukon River chum salmon — the "summer," and "fall" chums. In recent years state fishery biologists have learned surprising things about these two runs of chums. Although they are the same species, they differ greatly.

Summer chums arrive with king salmon at the mouth of the river in June, and they continue to enter the river until about mid-July. Their average weight is from 6½ to 7 pounds. Soon after entering fresh water they start to turn color, rapidly assuming the red, yellow, brown "calico" appearance typical of spawning chums long in fresh water. They darken quickly, and their flesh starts to bleach, becoming increasingly paler as their life nears an end. Eventually they become almost black, of poor flesh, and, of course, unmarketable, except for their roe, which is of high quality in the upper Yukon area fishing districts.

These summer chums spawn generally in tributaries of the lower 500 miles of the Yukon River in a wide diversity of river types, from tiny creeks to large rivers. In some years biologists estimate there may be as many as 5 million summer chums in the Yukon. Runs in recent years have ranged from about 1.3 to 2.6 million fish, a rough estimate based on tag-recovery information gathered in 1970-71.

Fishery biologists have counted more than 1 million summer chums

spawning in the Anvik River, a 140-mile-long tributary that flows into the west side of the lower Yukon.

Fall chum salmon appear in the lower Yukon in mid-July, and their run peaks in August, although they may still enter the Yukon as late as the end of September. They are more robust than the summer chums, averaging about 8 pounds. In contrast to summer chums, fall chums remain bright silver, almost appearing ocean-fresh even when they are more than several hundred miles from salt water. They do become water-marked or calico-colored as they reach their far-upriver spawning grounds.

The flesh of fall chums remains firm and with good color for most of the time they are in the main Yukon, making them a highly desirable fish, comparable in value to the finest chums taken in salt water. Locals often speak of them as "silver salmon," a reference to their shiny appearance. This causes confusion, for one of the common names of the coho salmon is also "silver salmon": and small numbers of the true coho or silver salmon are also in the Yukon during the fall.

Fall chums spawn mostly in far upriver streams (their spawning has been documented in 32 Alaska and 6 Yukon Territory tributaries to the Yukon) and the areas they seek for spawning are peculiar to the fall race or run of chums: they spawn in upwelling spring areas, where water temperatures range from 34° F to 40° F.

Air temperatures may drop to -40°

or -50° surrounding these spring water spawning pools, but the water where these fish spawn remains unfrozen. Several of the unique spawning areas for fall chums are glacial streams during summer, with silty water: in the cool fall glacial action stops, leaving the clear upwelling springs.

Major streams where fall chums spawn include the Fishing Branch River of the upper Porcupine River, in Yukon Territory, 1,600 miles from the mouth of the Yukon, where 350,000 fall chums were once counted in a 10-mile stretch; the mainstream of the Tanana River above the Richardson Highway bridge near Delta Junction; the lower

Most Yukon River drainage fish wheels have two baskets. This unusual one, on the Tanana River near Nenana, has three baskets — quite an engineering feat. (Len Sherwin, reprinted from ALASKA® magazine)

mile of the Delta River; the Toklat River on the north side of the Alaska Range; and the Sheenjek, a tributary of the Porcupine River in Alaska.

Ron Regnart, Alaska Department of Fish and Game regional supervisor for the Arctic-Yukon-Kuskokwim region, has watched late November spawning of fall chums. "Air temperature was below zero. Backs of the fish were sticking out of water as the fish spawned in mere inches of water. I could easily pick them up, for they were sluggish. Above the waterline they were frozen stiff, although their bellies were still soft.

"I could pick them out of the shallows and hold them and scrape ice and frost off their frozen backs. Their dorsal fins were frozen and partly broken off. Those fish were literally half-frozen. I put them back into deep water and watched them swim about, thawing out."

These unusual chums behave differently from other salmon when they enter the mouth of the Yukon. They arrive in huge schools in a run that may last for two days, or half a day. And then there are virtually no fish in the river until the next bunch arrives. There may be three or four of these surges of fall chums during the approximately month-long run. Runs, when they arrive, are commonly very heavy. "It's like turning a faucet on and off," says Regnart.

Tagging has shown that the majority of the first arriving fall chums tend to go into the Porcupine River drainage. Two or three weeks later the majority

of chums tends to go into the Tanana River drainage.

In the upper Yukon there is not only a timing difference between the early and late fall chums, there is a spatial difference where they run. Chums that swim into the Porcupine River to eventually spawn in the Fishing Branch River in Yukon Territory, and the Sheenjek in Alaska, and probably others along the Porcupine, swim along the north bank of the Yukon, while chums bound for the Tanana, Delta, and Toklat rivers swim mostly along the south bank.

Separation of these two basic stocks occurs as far downriver as Galena, where the great Yukon is about three-fourths of a mile wide.

Since the fishing equipment of the upper river is fixed (doesn't move about — fishes one location) fishery managers can selectively harvest fish bound for these two areas, and they do: the upper Yukon River is split into two fishery management units, *with the boundary drawn down the middle of the river.*

Coho salmon are relatively unimportant on the Yukon River to both commercial and subsistence fishermen. The annual average catch 1973-82 was 20,400. Cohos enter the Yukon in late July, about a week after arrival of the fall chum salmon, and fishermen who set their nets or fish wheels for chums incidentally catch the 7-pound-average silvers.

Pink and sockeye salmon are found in limited numbers in the lower Yukon; neither is found in commercially significant numbers.

Salmon catch for
Yukon River Fishing District

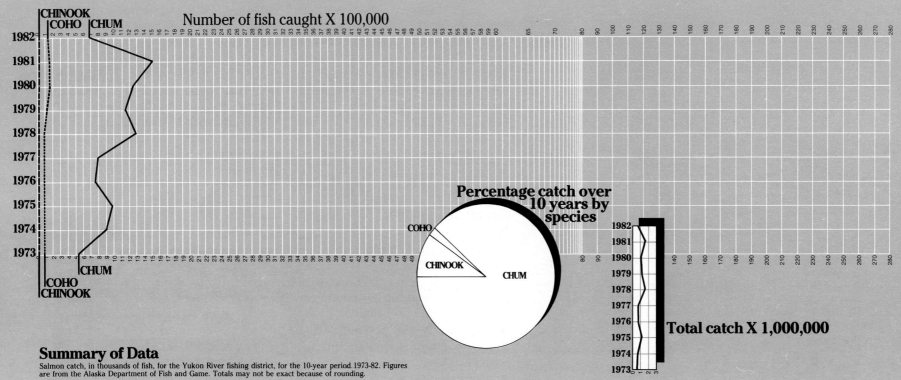

Number of fish caught X 100,000

Percentage catch over 10 years by species

Total catch X 1,000,000

Summary of Data

Salmon catch, in thousands of fish, for the Yukon River fishing district, for the 10-year period 1973-82. Figures are from the Alaska Department of Fish and Game. Totals may not be exact because of rounding.

YEAR	CHINOOK	SOCKEYE	COHO	PINK	CHUM	TOTAL
1982	121.7	—	29.0	—	661.7	812.4
1981	158.1	—	23.7	—	1,473.1	1,655.1
1980	154.0	—	8.7	—	1,222.0	1,384.8
1979	127.7	—	17.2	—	1,139.3	1,284.2
1978	99.2	—	26.2	—	1,288.6	1,413.0
1977	96.8	—	38.9	—	792.9	928.5
1976	88.3	—	5.2	—	756.8	850.3
1975	63.9	—	2.5	—	985.3	1,051.8
1974	97.9	—	16.2	—	879.2	993.4
1973	75.4	—	36.6	—	517.9	629.9
TOTAL	1,083.0	—	204.2	.4	9,716.8	11,004.3
AVERAGE	108.3	—	20.4	—	971.6	1,100.4
% Each Species 10 Yrs.	9.8%	—	1.8%	—	88.3%	
% Of Statewide Catch For Species, 10 Yrs.	16.0%	—	.7%	—	13.3%	ALL: 1.6%

Limited entry permits held for the upper Yukon fishing district (September 1982) included 166 for fish wheels, and 76 for gill nets: in addition there were 710 gill net entry permits for the lower Yukon.

Henry Oyoumick returning to Unalakleet after picking his gill net.
(Susan Hackley Johnson, reprinted from ALASKA GEOGRAPHIC®)

Norton Sound

A minor salmon fishery with its center about 175 miles south of the Arctic Circle, the **Norton Sound** salmon fishery brings badly needed dollars to the 203 mostly Eskimo gill-net fishermen who hold permits to fish there. The annual catch 1973-82 averaged 346,500 salmon, of which 47.6% (165,100) were chums, 44.4% (154,100) were pinks, and 6.2% (21,700) were coho. An average annual catch of 5,400 kings, representing 1.5% of the total number of salmon caught in Norton Sound were also taken in Norton Sound.

King salmon are uncommon north of the Shaktoolik River in Norton Sound, but have been found as far north as the Wulik River, about 100 miles northwest of Kotzebue. Norton Sound is the farthest north mixed species salmon fishery in North America: Kotzebue Sound's fishery, on and above the Arctic Circle 180 miles to the north of Norton Sound, is almost exclusively for a single species — chums.

The total catch in Norton Sound is less than 1% of the statewide total.

Best catch on record for Norton Sound was in 1982, when just over half a million (511,200) salmon were netted. Of these 230,300 were pink salmon, and that was just a fraction of the pink salmon that were available for harvest. The problem? Few buyers.

The pink salmon of Norton Sound swarm back by the millions in some years. In 1982 upward of 10 million pinks returned to Norton Sound, reported state fishery managers. "The largest pink run recorded since statehood," they said. The sonar salmon counter in the Unalakleet River, one of the area's biggest pink producers, counted 6.1 million pinks. Other major producing salmon systems in Norton Sound include the Niukluk, Fish, Kwiniuk, Tubutulik, Shaktoolik, Koyuk, Ungalik and Inglutalik rivers.

What's wrong with Norton Sound pink salmon? They're small — averaging about 3 pounds. The small domestic demand resulted in the sale of 230,000 of the diminutive salmon in 1982. Foreign buyers were invited to Norton Sound that year and told they could buy the pink salmon, but none arrived.

No road connects this area with the rest of the state. Nome has a small network of roads surrounding it, and other commerce centers have short lengths of road. Air transportation is the primary way of getting to and from, and around, this region. Normal average summer temperatures range from 30° F to 50° F and normal average winter temperatures from 5° F to 10° F. Precipitation ranges from 15 to 20 inches a year.

About 10,000 people, mostly Eskimos, live in the Norton Sound-Kotzebue Sound region. Communities on the shores of Norton Sound include the old gold rush town of Nome (pop. 2,273), Unalakleet (615), Golovin (94), Shaktoolik (177), and Koyuk (203).

In 1982, about 99,000 salmon were caught by subsistence fishermen in Norton Sound, using gill nets.

Salmon catch for
Norton Sound

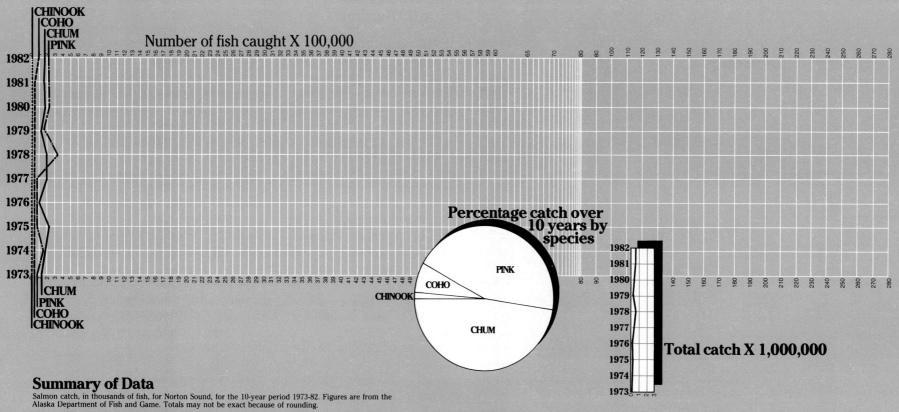

Number of fish caught X 100,000

Percentage catch over 10 years by species

PINK
COHO
CHINOOK
CHUM

Total catch X 1,000,000

Summary of Data

Salmon catch, in thousands of fish, for Norton Sound, for the 10-year period 1973-82. Figures are from the Alaska Department of Fish and Game. Totals may not be exact because of rounding.

YEAR	CHINOOK	SOCKEYE	COHO	PINK	CHUM	TOTAL
1982	5.9	—	91.6	230.3	183.4	511.2
1981	7.9	.1	31.2	229.6	169.8	438.6
1980	6.3	—	30.0	227.2	180.7	444.3
1979	10.8	.2	31.1	164.9	138.0	345.0
1978	10.0	—	7.3	325.6	189.2	532.1
1977	4.5	.2	3.7	48.7	200.5	257.3
1976	2.2	—	6.9	87.9	96.0	193.2
1975	2.4	—	4.6	32.4	212.5	251.9
1974	3.0	—	2.1	148.2	162.2	315.4
1973	1.9	—	9.3	46.5	119.1	176.8
TOTAL	54.9	.5	217.8	1,541.3	1,651.4	3,465.8
AVERAGE	5.4	—	21.7	154.1	165.1	346.5
% Each Species 10 Yrs.	1.5%	—	6.2%	44.4%	47.6%	
% Of Statewide Catch For Species, 10 Yrs.	0.8%	—	.8%	.4%	2.2%	ALL: 5%

Limited entry permits held for Norton Sound: 203 for gill nets.

The Kotzebue Sound District

The **Kotzebue Sound** salmon fishing district is similar to that of Norton Sound in that the catch is relatively small, producing during the years 1973-82 an annual average catch of 364,700 salmon, 99.7% of which were chums. An average of about 1,340 pinks were also netted here by the 223 mostly Eskimo commercial fishermen who hold entry permits for gill nets for this fishery.

Best catch for the 10-year period was in 1981, when 677,400 of the 9-pound-average chums were netted. This is the farthest north commercial Pacific salmon fishery in North America: north of the Kotzebue Sound drainage all five species of Pacific salmon become relatively scarce.

Commercial salmon fishing in the Kotzebue district dates back to 1914-18 when 10,130 cases and 300 barrels of hard salt salmon were processed near Kotzebue.

The next commercial effort was in 1962, and since then a consistent harvest has been made. Average harvest until the mid-1970s was around 179,000 fish annually. Fishing effort and interest increased rapidly to the present level.

Major salmon streams in the Kotzebue Sound district include the Kivalina, Wulik, Noatak, and Kobuk.

Kotzebue (pop. 2,044), Noorvik (490), Selawik (361), Point Hope (461), and Kiana (344) are important communities in Kotzebue Sound.

Subsistence fishing for salmon is important to some residents of Kotzebue Sound. In 1982 at least 33,000 salmon were taken by local fishermen there.

*Above — **A commercial fisherman sells his chum salmon catch at Kotzebue. Even though Kotzebue Sound is the northern limit of their range, chums from this fishery bring high prices because of their high quality and because their run culminates when other salmon runs along the North Pacific coast are at a low ebb.***
(Dave Sweigert, reprinted from ALASKA GEOGRAPHIC®)

*Right — **Commercial fishermen check a set net curved into an S shape by the changing tide of Kotzeube Sound. In the foreground grows ryegrass, typical of beach vegetation in this area.***
(Anore Jones, reprinted from ALASKA GEOGRAPHIC®)

Salmon catch for
Kotzebue Sound

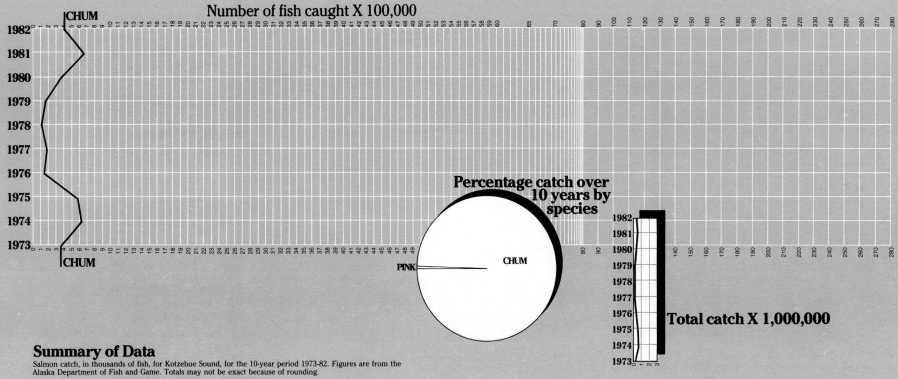

Number of fish caught X 100,000

Percentage catch over 10 years by species

Total catch X 1,000,000

Summary of Data

Salmon catch, in thousands of fish, for Kotzebue Sound, for the 10-year period 1973-82. Figures are from the Alaska Department of Fish and Game. Totals may not be exact because of rounding.

YEAR	CHINOOK	SOCKEYE	COHO	PINK	CHUM	TOTAL
1982	—	—	—	—	417.7	417.7
1981	.1	.1	—	.2	677.4	677.7
1980	.2	—	—	1.5	366.5	368.2
1979	.2	—	—	.7	141.5	142.5
1978	.1	—	—	7.0	111.6	118.7
1977	—	—	—	—	192.5	192.5
1976	—	—	—	—	159.7	159.7
1975	—	—	—	—	563.7	563.7
1974	—	—	—	—	628.0	628.1
1973	—	—	—	—	378.6	378.6
TOTAL	.6	.1	—	9.4	3,637.2	3,647.4
AVERAGE	—	—	—	.9	363.7	364.7
% Each Species 10 Yrs.	—	—	—	.02%	99.7%	
% Of Statewide Catch For Species, 10 Yrs.	—	—	—	—	4.9%	ALL: 05%

There are 223 limited entry permits for gill nets for Kotzebue Sound.

Alaska Geographic® Back Issues

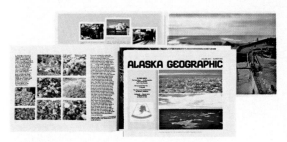

The North Slope, Vol. 1, No. 1. The charter issue of *ALASKA GEOGRAPHIC*® took a long, hard look at the North Slope and the then-new petroleum development at "the top of the world." *Out of print.*

Fisheries of the North Pacific: History, Species, Gear & Processes, Vol. 1, No. 4. The title says it all. This volume is out of print, but the book, from which it was excerpted, is available in a revised, expanded large-format volume. 424 pages. $24.95.

Prince William Sound, Vol. 2, No. 3. This volume explored the people and resources of the Sound. *Out of print.*

One Man's Wilderness, Vol. 1, No. 2. The story of a dream shared by many, fulfilled by few: a man goes into the bush, builds a cabin and shares his incredible wilderness experience. Color photos. 116 pages, $9.95

The Alaska-Yukon Wild Flowers Guide, Vol. 2, No. 1. First Northland flower book with both large, color photos and detailed drawings of every species described. Features 160 species, common and scientific names and growing height. Vertical-format book edition now available. 218 pages, $12.95.

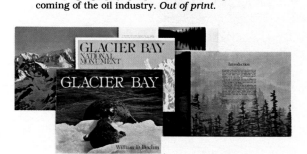

Yakutat: The Turbulent Crescent, Vol. 2, No. 4. History, geography, people — and the impact of the coming of the oil industry. *Out of print.*

Admiralty . . . Island in Contention, Vol. 1, No. 3. An intimate and multifaceted view of Admiralty: its geological and historical past, its present-day geography, wildlife and sparse human population. Color photos. 78 pages, $5.00

Richard Harrington's Yukon, Vol. 2, No. 2. The Canadian province with the colorful past *and* present. *Out of print.*

Glacier Bay: Old Ice, New Land, Vol. 3, No. 1. The expansive wilderness of Southeastern Alaska's Glacier Bay National Monument (recently proclaimed a national park and preserve) unfolds in crisp text and color photographs. Records the flora and fauna of the area, its natural history, with hike and cruise information, plus a large-scale color map. 132 pages, $9.95

The Land: Eye of the Storm, Vol. 3, No. 2. The future of one of the earth's biggest pieces of real estate! *This volume is out of print,* but the latest on the Alaska lands controversy is detailed completely in Volume 8, Number 4.

Richard Harrington's Antarctic, Vol. 3, No. 3. The Canadian photojournalist guides readers through remote and little understood regions of the Antarctic and Subantarctic. More than 200 color photos and a large fold-out map. 104 pages, $8.95

The Silver Years of the Alaska Canned Salmon Industry: An Album of Historical Photos, Vol. 3, No. 4. The grand and glorious past of the Alaska canned salmon industry. *Out of print.*

Alaska's Volcanoes: Northern Link in the Ring of Fire, Vol. 4, No. 1. Scientific overview supplemented with eyewitness accounts of Alaska's historic volcano eruptions. Includes color and black-and-white photos and a schematic description of the effects of plate movement upon volcanic activity. 88 pages. *Temporarily out of print.*

The Brooks Range: Environmental Watershed, Vol. 4, No. 2. An impressive work on a truly impressive piece of Alaska — The Brooks Range. *Out of print.*

Kodiak: Island of Change, Vol. 4, No. 3. Russians, wildlife, logging and even petroleum . . . an island where change is one of the few constants. *Out of print.*

Wilderness Proposals: Which Way for Alaska's Lands?, Vol. 4, No. 4. This volume gave yet another detailed analysis of the many Alaska lands questions. *Out of print.*

Cook Inlet Country, Vol. 5, No. 1. Our first comprehensive look at the area. A visual tour of the region — its communities, big and small, and its countryside. Begins at the southern tip of the Kenai Peninsula, circles Turnagain Arm and Knik Arm for a close-up view of Anchorage, and visits the Matanuska and Susitna valleys and the wild, west side of the inlet. *Out of print.*

Southeast: Alaska's Panhandle, Vol. 5, No. 2. Explores Southeastern Alaska's maze of fjords and islands, mossy forests and glacier-draped mountains — from Dixon Entrance to Icy Bay, including all of the state's fabled Inside Passage. Along the way are profiles of every town, together with a look at the region's history, economy, people, attractions and future. Includes large fold-out map and seven area maps. 192 pages, $12.95.

Bristol Bay Basin, Vol. 5, No. 3. Explores the land and the people of the region known to many as the commercial salmon-fishing capital of Alaska. Illustrated with contemporary color and historic black-and-white photos. Includes a large fold-out map of the region. *Out of print.*

Alaska Whales and Whaling, Vol. 5, No. 4. The wonders of whales in Alaska — their life cycles, travels and travails — are examined, with an authoritative history of commercial and subsistence whaling in the North. Includes a fold-out poster of 14 major whale species in Alaska in perspective, color photos and illustrations, with historical photos and line drawings. 144 pages, $12.95.

Yukon-Kuskokwim Delta, Vol. 6, No. 1. This volume explored the people and lifestyles of one of the most remote areas of the 49th state. *Out of print.*

The Aurora Borealis, Vol. 6, No. 2. Here one of the world's leading experts — Dr. S.-I. Akasofu of the University of Alaska — explains in an easily understood manner, aided by many diagrams and spectacular color and black-and-white photos, what causes the aurora, how it works, how and why scientists are studying it today and its implications for our future. 96 pages, $7.95.

Alaska's Native People, Vol. 6, No. 3. In this edition the editors examine the varied worlds of the Inupiat Eskimo, Yup'ik Eskimo, Athabascan, Aleut, Tlingit, Haida and Tsimshian. Included are sensitive, informative articles by Native writers, plus a large, four-color map detailing the Native villages and defining the language areas. 304 pages, $24.95.

The Stikine, Vol. 6, No 4. River route to three Canadian gold strikes in the 1800s. This edition explores 400 miles of Stikine wilderness, recounts the river's paddlewheel past and looks into the future. Illustrated with contemporary color photos and historic black-and-white; includes a large fold-out map. 96 pages, $9.95.

Alaska's Great Interior, Vol. 7, No. 1. Alaska's rich Interior country, west from the Alaska-Yukon Territory border and including the huge drainage between the Alaska Range and the Brooks Range, is covered thoroughly. Included are the region's people, communities, history, economy, wilderness areas and wildlife. Illustrated with contemporary color and black-and-white photos. Includes a large fold-out map. 128 pages, $9.95.

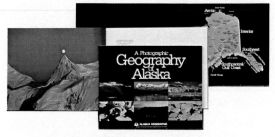

A Photographic Geography of Alaska, Vol. 7, No. 2. An overview of the entire state — a visual tour through the six regions of Alaska: Southeast, Southcentral/Gulf Coast, Alaska Peninsula and Aleutians, Bering Sea Coast, Arctic and Interior. Plus a handy appendix of valuable information — "Facts About Alaska." Approximately 160 color and black-and-white photos and 35 maps. 192 pages, $14.95.

The Aleutians, Vol. 7, No. 3. Home of the Aleut, a tremendous wildlife spectacle, a major World War II battleground and now the heart of a thriving new commercial fishing industry. Contemporary color and black-and-white photographs, and a large fold-out map. 224 pages, $14.95.

Alaska Mammals, Vol. 8, No. 2. From tiny ground squirrels to the powerful polar bear, and from the tundra hare to the magnificent whales inhabiting Alaska's waters, this volume includes 80 species of mammals found in Alaska. Included are beautiful color photographs and personal accounts of wildlife encounters. 184 pages, $12.95.

Alaska's Glaciers, Vol. 9, No. 1. Examines in-depth the massive rivers of ice, their composition, exploration, present-day distribution and scientific significance. Illustrated with many contemporary color and historical black-and-white photos, the text includes separate discussions of more than a dozen glacial regions. 144 pages, $9.95.

Klondike Lost: A Decade of Photographs by Kinsey & Kinsey, Vol. 7, No. 4. An album of rare photographs and all-new text about the lost Klondike boom town of Grand Forks, second in size only to Dawson during the gold rush. Introduction by noted historian Pierre Berton: 138 pages, area maps and more than 100 historical photos, most never before published. $12.95.

The Kotzebue Basin, Vol. 8, No. 3. Examines northwestern Alaska's thriving trading area of Kotzebue Sound and the Kobuk and Noatak river basins. Contemporary color and historical black-and-white photographs. 184 pages, $12.95.

Sitka and Its Ocean/Island World, Vol. 9, No. 2. From the elegant capital of Russian America to a beautiful but modern port, Sitka, on Baranof Island, has become a commercial and cultural center for Southeastern Alaska. Pat Roppel, longtime Southeast resident and expert on the region's history, examines in detail the past and present of Sitka, Baranof Island, and neighboring Chichagof Island. Illustrated with contemporary color and historical black-and-white photographs. 128 pages, $9.95.

Wrangell-Saint Elias, Vol. 8, No. 1. Mountains, including the continent's second- and fourth-highest peaks, dominate this international wilderness that sweeps from the Wrangell Mountains in Alaska to the southern Saint Elias range in Canada. Illustrated with contemporary color and historical black-and-white photographs. Includes a large fold-out map. 144 pages, $9.95.

Alaska National Interest Lands, Vol. 8, No. 4. Following passage of the bill formalizing Alaska's national interest land selections (d-2 lands), longtime Alaskans Celia Hunter and Ginny Wood review each selection, outlining location, size, access, and briefly describing the region's special attractions. Illustrated with contemporary color photographs. 242 pages, $14.95.

Islands of the Seals: The Pribilofs, Vol. 9, No. 3.
Great herds of northern fur seals drew Russians and Aleuts to these remote Bering Sea islands where they founded permanent communities and established a unique international commerce. The communities languished under U.S. control until recent decades when new legislation and attempts at economic diversification have increased interest in the islands, their Aleut people, and the rich marine resources nearby. Illustrated with contemporary color and historical black-and-white photographs. 128 pages, $9.95.

Adventure Roads North: The Story of the Alaska Highway and Other Roads in *The MILEPOST* ®, Vol. 10, No. 1. From Alaska's first highway — the Richardson — to the famous Alaska Highway, first overland route to the 49th state, text and photos provide a history of Alaska's roads and take a mile-by-mile look at the country they cross. 224 pages, $14.95.

The Alaska Geographic Society

Box 4-EEE, Anchorage, AK 99509

Membership in The Alaska Geographic Society is $30, which includes the following year's four quarterlies which explore a wide variety of subjects in the Northland, each issue an adventure in great photos, maps, and excellent research. Members receive their quarterlies as part of the membership fee at considerable savings over the prices which nonmembers must pay for individual book editions.

Alaska's Oil/Gas & Minerals Industry, Vol. 9, No. 4. Experts detail the geological processes and resulting mineral and fossil fuel resources that are now in the forefront of Alaska's economy. Discussions of historical methods and the latest techniques in present-day mining, submarine deposits, taxes, regulations, and education complete this overview of an important state industry. Illustrated with historical black-and-white and contemporary color photographs. 216 pages, $12.95.

ANCHORAGE and the Cook Inlet Basin . . . Alaska's Commercial Heartland, Vol. 10, No. 2. An update of what's going on in "Anchorage country" . . . the Kenai, the Susitna Valley, and Matanuska. Heavily illustrated in color and including three illustrated maps . . . one an uproarious artist's forecast of "Anchorage 2035." 168 pages, $14.95.

NEXT ISSUE
Koyukuk Country, Vol. 10, No. 4. This issue explores the vast drainage of the Koyukuk River, third largest in Alaska. Text and photos provide information on the land and offer insights into the lifestyle of the people who live and have lived along the Koyukuk. To members in November 1983. Price to be announced.